PROJECT AIR FORCE

T0127925

Increasing Cost-Effective Readiness for the U.S. Air Force by Reducing Supply Chain Variance

Technical Analysis of Flying Hour Program Variance

Patrick Mills, Sarah A. Nowak, Peter Buryk, John G. Drew, Christopher Guo, Raffaele Vardavas

Prepared for the United States Air Force

For more information on this publication, visit www.rand.org/t/RR2118

Library of Congress Control Number: 2018938566

ISBN: 978-0-8330-9885-6

Published by the RAND Corporation, Santa Monica, Calif.

© Copyright 2018 RAND Corporation

RAND® is a registered trademark.

Support RAND

Make a tax-deductible charitable contribution at

www.rand.org/giving/contribute

www.rand.org

Preface

The U.S. Air Force spends considerable funds to operate and support its aircraft. Between fiscal years (FYs) 1996 and 2011, such spending increased by more than 6 percent a year, limiting what could be spent on other priorities. One way to reduce operation and support costs is to improve the accuracy of demand forecasts. The Air Force spends about $4 billion annually to buy and repair spare parts for aircraft. Demand that runs lower than forecast levels results in excess parts; demand that runs higher results in shortages and reduced readiness. One potential way to improve spare part demand forecasts is to reduce the difference between the number of flying hours that are forecast and the number that are actually flown, which is known as flying hour variance. The Air Force Sustainment Center asked RAND Project AIR FORCE to gauge the potential effect of flying hour variance on cost and readiness, identify the causes of the variance and quantify their effects, and identify policy options to rectify problems identified.

The research reported here was commissioned by the Air Force Sustainment Center and conducted within the Resource Management Program of RAND Project AIR FORCE as part of an FY 2014 project Cost-Effective Readiness. This happened to be the same time that some of the worst effects of the budget sequester were being felt. Most of the data presented in this report originally spanned FYs 2008–2012 because those dates were the most-widely available in FY 2014 when the research was conducted. However, after the research was completed and the document reviewed, the project sponsor felt that part of the story was missing because the report essentially ignored the sequester. In the interest of telling a more complete story, more data were obtained, up to FY 2015, to include the more significant perturbations from such events as the sequester and the averted A-10 retirement. All of the analyses and figures whose conclusions hinge on those specific years have been updated.

RAND Project AIR FORCE

RAND Project AIR FORCE (PAF), a division of the RAND Corporation, is the U.S. Air Force's federally funded research and development center for studies and analyses. PAF provides the Air Force with independent analyses of policy alternatives affecting the development, employment, combat readiness, and support of current and future air, space, and cyber forces. Research is conducted in four programs: Force Modernization and Employment; Manpower, Personnel, and Training; Resource Management; and Strategy and Doctrine. The research reported here was prepared under contract FA7014-16-D-1000.

Additional information about PAF is available on our website: http://www.rand.org/paf.

This report documents work originally shared with the U.S. Air Force on September 23, 2014. The draft report, issued on February 26, 2016, was reviewed by formal peer reviewers and U.S. Air Force subject-matter experts.

Contents

Figures and Tables

Figures

Tables

Summary

The Air Force devotes considerable resources to aircraft operating and support, and spending in these areas increased annually by more than 6 percent between fiscal years (FYs) 1996 and 2011, before starting to level off. Such spending can have an outsized effect on the total budget and crowd out funding for other needs, such as modernization, readiness, and infrastructure. In response to this trend, the Air Force developed a new approach to balancing cost and effectiveness, which it labeled Cost-Effective Readiness (CER). CER aims to reduce the cost of each unit of readiness and has been adopted as one aim of the Air Force's Enterprise Logistics Strategy.

One area of emphasis for the strategy is to improve the accuracy of forecasts for spare parts. Several factors contribute to forecasting errors: natural uncertainty in spare part failures, flying hour (FH) variance, issues with phasing systems in and out of the Air Force, new items, and data errors. Many of these factors fall within the purview (though not necessarily the control) of the Air Force Materiel Command (AFMC).

One factor that falls outside the AFMC's purview is FH program (FHP) variance. The Air Force's FHP comprises the number of hours needed to attain and maintain combat readiness and capability for its aircrews, to test weapon systems and tactics, and to meet collateral requirements, such as air shows, demonstration rides for important personnel, and ferrying aircraft. Operations and training personnel at the major commands (MAJCOMs) determine the number of hours required annually.[1]

Each year, the Air Force estimates how many hours will be flown for pilot training sorties and then uses these estimates to forecast its demand for depot-level reparables. These forecasts drive a number of other logistics decisions and processes. When the actual number of hours flown differs from the number forecast, the difference can affect readiness or cause more spending than is needed (e.g., buying spare parts in excess of near-term demand), which prevents the money from being used for other important priorities. The Air Force Sustainment Center (AFSC) asked RAND Project AIR FORCE to assess the potential effect of FHP variance on cost and readiness. This report responds to that request.

As part of the annual planning, programming, budgeting, and execution system process, also referred to as the program objective memorandum (POM) process, the air staff, with inputs from the MAJCOMs, determines the annual FHP. Funding for the FHP is subject to the perturbations of the overall POM process. Sometimes FHP funding is reduced to make room for other requirements. Sometimes those cuts are then restored in the year of execution as funding

---

[1] U.S. Air Force, Air Force Instruction 11-102, *Flying Hour Program Management*, August 30, 2011.

continues to shift, sending contradictory signals to the supply chain community.[2] Nonetheless, an FHP is eventually established, setting off a series of further decisions and outcomes.

Planned FHs—taken directly from the FHP—drive the demand forecast. This, in turn, drives decisions to buy spare parts and set depot repair capacity—essentially organic depot labor levels—in the form of annual contracts with unionized government civilian personnel. At some point, because of lead times, those decisions cannot be reversed without additional investment of time, trouble, money, or some combination thereof.

The FHP recorded in the budget influences what is actually flown, but many external factors can influence what is actually flown in a given time period or over the entire year (e.g., fiscal changes, contingency operations, or safety issues, such as the grounding of a fleet). The operators setting the FHP in the POM process can conduct their process rigorously, but their product is still subject to those external factors. Then, downstream effects are subject to uncertainty, i.e., the stochastic relationship between flying activity and spare part breaks, between part breaks and spare part stock requests and repairs, and so on.[3] Many external factors disrupt the relationship between the FHP and actual downstream effects, including actions and decisions on the part of managers and maintainers to solve problems that are sometimes driven by inaccurate forecasts upstream.

Thus, because of the way the system is designed, the linkage between the FHP and downstream *decisions* is direct and straightforward, whereas the linkage between the FHP and downstream *activity* is more tenuous and is governed by a range of factors, each with inherent uncertainty.

To better understand the dynamics of FH variance, we analyzed FH data for aircraft and individual spare parts. We found that, in both cases, those aircraft or parts with the highest numbers of actual FHs generally had low statistical error, and the highest statistical error was from those aircraft and parts with relatively few hours flown. We then summed up the total error for low- and high-error aircraft and parts. We found that, in both cases (aircraft and parts), the majority of total error was driven by a large number of aircraft or parts with small individual error. What this means is that, although high-error aircraft and parts can be big drivers in their individual programs, they do not significantly affect the total error in the system. Instead, it is the accumulation of many, many small errors that drives most of the total FH forecast error.

We group the range of causes of FHP variance into three broad categories. Simple planning error accounts for the basic uncertainties in predicting the FHP in a given year, including the

---

[2] The CER working group highlighted a number of instances of FHP changes that created significant volatility: in FY 2008, a 10-percent FH cut; in FY 2009, the Combat Air Forces aircraft reduction; in FY 2011, the FHP optimized and reduced because of overseas contingency operations; in FY 2012, a 5-percent Combat Air Forces FHP efficiency; and in FY 2013, the budget sequester.

[3] At a very high level, e.g., all parts for a single aircraft type, FHs tend to correlate reasonably well with spare part failures. At more-disaggregated levels, there is generally low correlation between the two, i.e., a statistically "noisy" relationship is displayed.

number of pilots, number and type of sorties, and sortie duration. External causes are those that originate outside the service and that usually affect the entire enterprise or a significant portion of it, such as contingency operations or congressional action. The third category is internal Air Force decisions, which can cause FHP variance when far-reaching decisions (about force structure, budgets, or FHs themselves) are made after the original FHP is set. Understanding these various sources is key to crafting policy solutions to address and reduce FHP variance.

We analyzed the enterprise-level effect of FHP variance on several downstream metrics. FHP variance—regardless of its source—increases forecast error, the source of all other downstream effects. Underflying (i.e., overplanning) can incur opportunity cost, leaving money on the table in the budget process. It also incurs financial costs in the form of holding costs for unneeded inventory. Overflying (i.e., underplanning) likely contributes to readiness problems, but our analysis found no statistically significant relationship between FHP variance and mission impaired capability awaiting parts incidents, one of several important aircraft readiness metrics.

However, each effect described above comes with caveats. Though forecast error induced by FHP variance for specific aircraft fleets can be enormous, in most years, the effect on enterprise-level demand forecast accuracy (DFA) was modest. Except for one year of sequestration, we found the average enterprise-level increase to forecast error to be about five points, or an increase of about 15 percent over baseline error.

Budget opportunity cost can be high—hundreds of millions of dollars in a single year—but many of the recent sources of volatility were events for which flying funding itself was cut from the budget after the FHP was set. Thus, the FH budget was not necessarily too large (i.e., leaving money on the table), even though hours were underflown from their original estimate.

Financial costs incurred from underflying appear to be low, about $2 million to $4 million per year for inventory holding costs in recent years, including the years of sequestration.

In sum, we found that, at an enterprise level, the one outcome with the most direct connection to FHs—budget opportunity cost—has a direct and significant link to FHP variance. But in recent years, the enterprise-level effect on the other three—DFA, financial cost, and readiness—are more tenuous. Thus, because the Air Force pursues enterprise-level improvements to cost-effectiveness, it might want to look beyond FHP variance. We make one recommendation below on how to do that.

However, individual programs do sometimes have extremely high FH variance and thus do experience larger downstream effects. The AFSC's planning processes do eventually catch large perturbations in FHP inputs, but that does not preclude supply chain planners and operators from having to respond to them. AFSC subject-matter experts (SMEs) reported that, on several occasions, they were caught off guard by seemingly sudden, radical changes to an aircraft's FH forecast, with little or no communication from planners as to why and with little opportunity to communicate the downstream effects (e.g., canceled contracts, reduced or eliminated repair capacity). This no doubt drives a portion of the AFSC's concern with the accuracy of FHP forecasts.

Some steps have already been taken that could address this gap in communication (at least in part spurred by increased scrutiny from CER). For example, the Deputy Chief of Staff for Operations, Plans, and Requirements (AF/A3) issued a memorandum to increase communication and coordination among Headquarters (HQ), Air Force, organizations involved in the FHP.[4] Besides general emphasis on synchronization and awareness, the memo states that MAJCOMs must explain under- or overexecution and notify HQ AFSC (among other organizations) of approved FH realignment actions.

However, this apparently has not produced the desired results, and the integration of stakeholders involved in the FH processes has not been incorporated into efforts aimed at improving FH variance.[5] As a result, HQ AFSC has formed an eight-step Cross Command Flying Hour Program Working Group to continue and intensify efforts needed to improve integration and communication regarding the development of FH programs. This should help address extreme program-level perturbations—which appear to be one of AFSC's biggest concerns—that might cause undue FHP variance if supply chain planners are not kept in the loop.

The air staff made several other changes to planning processes: planning FHs at the mission design series (MDS) level, setting improvement targets for FH variance, and including a factor for deployments (which had, in some cases, been excluded).[6] These changes should be most successful at reducing the natural error in the FH planning process because they are aimed at the fundamental processes that produce the estimates.

The guiding question in this analysis is how to achieve CER. Our analysis suggests that there are two separate issues or concerns here. The first is about opportunity cost, essentially a question of developing a budget. The second is about demand forecasting and its downstream effects.

Significant overplanning can actually reduce budget tradespace by a significant amount that could be invested in other important programs. At an enterprise level, FHs are a reasonable predictor of spare part removals and an excellent predictor of fuel consumption.[7] Thus, improving FH planning could contribute to the accuracy of the POM and free up badly needed resources in cases where the driver of overplanning was not a belated cut to the FH budget itself. The two HQ, Air Force actions referenced above should help address this.

However, this progress does not necessarily influence the second issue, that of enterprise-level forecast accuracy. Even with a more accurate planning process (i.e., the number-crunching

---

[4] Giovanni K. Tuck, HQ AF/A3, "FY15 Flying Hour Program Execution Guidance," memorandum, Washington, D.C., October 8, 2014.

[5] Email communication with HQ AFSC personnel on December 19, 2016.

[6] Email communication with HQ AFSC personnel on October 15, 2014.

[7] FHs are not a perfect predictor of fuel consumption, even by aircraft type, because different sortie profiles can drive very different flying activities and thus fuel demands.

that informs the POM input), FHs remain subject to severe volatility—all of those features inherent in the U.S. Department of Defense's current budget system and those external events that cannot be anticipated. Given the tenuous nature of the relationships we observed, further efforts to reduce FHP variance might or might not have an *observable* effect on long-term financial cost or readiness because so many other sources of error affect the system. Particular spare parts could be affected more by FHP variance because their removals correlate more highly with flying activity, but any such effects would not be observed system-wide.

With that in mind, the current forecasting system has at least two problems. First, it uses FHs as a direct, linear input, whereas FHs are themselves volatile.[8] FHs are subject to the budget process, so the Air Force is pegging its prediction on an input variable that is ever-shifting based in part on strategy but more so on unforeseeable institutional factors beyond its control and subject to fiscal pressures and budget negotiations.

The second, and maybe more important feature, is that FHs are generally poor predictors of actual removals and repair demands. The individual part level is what matters most for supply chain cost and effectiveness, but, at the part level, there is virtually no correlation with FHs.[9] Even if FHs never changed from the original POM forecast, they would still be poor predictors and would give poor DFA and other associated effects. In sum, the Air Force has chosen to drive its parts forecasts for flying depot-level reparables primarily by a single variable that is notoriously volatile and demonstrably unreliable.

So, could depot-level reparable removals be better forecast *without* better FHP forecasts? It is beyond the scope of this report to describe a comprehensive approach to improve the Air Force's spare part forecasting system, though we believe that such an approach is needed. However, we did discuss several possibilities with analysts in the 448th Supply Chain Management Wing and Analysis Directorate, Strategic Plans, Programs, Requirements and Assessments Division, Air Force Materiel Command (AFMC/A9A).

The current forecasting system uses a calculated demand rate (removals per FH) and allows for human intervention when equipment specialists have additional applicable information about anticipated future demands (e.g., phasing in or out parts or aircraft). One possibility is, instead of using just a removal rate, to supplement or replace that with a time-based failure rate, such as demands per quarter. Separate analyses by the RAND Project AIR FORCE research team, 448th Supply Chain Management Wing, and AFMC/A9A have shown that DFA can improve when

---

[8] Manual overrides are used in cases of known or anticipated program changes, but data analysis shows that, in aggregate, these overrides generally increase total forecast error.

[9] Past RAND research has noted that one fundamental assumption underlying the spare part forecasting system is not supported by the data. In other words, the so-called "linearity assumption": "Aircraft failures are driven by a known operational activity: the expected number of failures of a particular part is proportional to a known and measurable quantity, such as FHs or landing." See Gordon B. Crawford, *Variability in the Demands for Aircraft Spare Parts: Its Magnitude and Implications*, Santa Monica, Calif.: RAND Corporation, R-3318, 1988.

historical removals are used instead of the current method of removal rates, either discounting FHs or ignoring them altogether.[10]

Another possibility is to reduce SME intervention in the removal forecast. We found that, whether using time-based or FH-based failure rates, on the whole, SME input actually worsened forecasts and reduced DFA. Human intervention is necessary in the case of known significant events (e.g., time change technical order modifications), but the data suggest that they should be used sparingly, i.e., when they can be shown to have a measurably positive effect.

Also, past RAND research has pointed to some potential solutions. Adams, Abell, and Isaacson (1993) lays out an approach to better forecasting high-demand items using a weighted regression technique.[11] And a number of RAND studies from the mid-1960s showed that sorties rather than FHs drove failures.[12]

Finally, HQ AFSC is working to implement a method called Peak Policy for low-demand, highly variable items. That methodology has been implemented by the Defense Logistics Agency for consumables, and AFSC is currently working to extend it to include reparables.

In light of these findings, we make four recommendations.

Maintain changes to the FHP planning process that appear to be essentially zero-cost to implement. These changes address mostly our first category of FHP variance, simple planning error, which, in recent years, has driven the majority of the overall volume of enterprise-level FH error. In addition to providing opportunity cost savings, addressing FHP variance should better balance cost and readiness across Air Force fleets. Resolving recent levels of FHP variance means that the Air Force would not overinvest in one fleet relative to another fleet.[13] However, reliably reducing the cost per unit of readiness (the goal of CER) requires that forecast error be reduced much more significantly than reducing FHP variance alone can accomplish.

Second, **continue to support and extend efforts to improve integration and communication** across commands and between the operational and supply chain communities, such as the Cross Command Flying Hour Program Working Group started by Logistics

---

[10] Some high-demand parts do actually show a reliable relationship between FHs and removals, so the D200 default could be retained. (D200 is a subsystem of the Air Force Requirements Management System.) A system that promised even better results was proposed in John L. Adams, John B. Abell, and Karen E. Isaacson, *Modeling and Forecasting the Demand for Aircraft Recoverable Spare Parts*, Santa Monica, Calif.: RAND Corporation, R-4211-AF/OSD, 1993.

[11] Adams, Abell, and Isaacson, 1993; John B. Abell, *Estimating Requirements for Aircraft Recoverable Spares and Depot Repair: Executive Summary*, Santa Monica, Calif.: RAND Corporation, R-4215-AF, 1993.

[12] For example, RAND research by William H. McGlothlin, Theodore S. Donaldson, and A. F. Sweetland, as well as Peter J. Francis and Geoffrey B. Shaw, *Effect of Aircraft Age on Maintenance Costs*, Alexandria, Va.: Center for Naval Analyses, CAB D0000289.A2, March 2000.

[13] One can imagine a cynical planner deliberately overestimating FHs for a particular MDS and, in the year of execution, simply achieving higher mission capability rates or fuller spares kits when those FHs do not materialize, to the detriment of other MDS that were more conservative in their planning.

Directorate, Air Force Sustainment Center (AFSC/LG). This type of effort seems to be the best hope to address our third category of FHP variance, internal Air Force decisions.

Third, **consider management mechanisms that could dampen the downstream volatility caused by FHP variance**. One approach to this would give supply chain managers more flexibility in responding to upstream changes in forecast data. There could be cases where a change in demands causes a downstream decision to cross a threshold (e.g., letting or canceling a contract). Such a decision, based on communication with upstream planners, might be delayed or forgone, especially if the demand forecast driving the decision departed significantly from historical demands. This type of thinking could be incorporated in data systems in a more automated way (e.g., updating demand rates and resupply times only when there has been a statistically supportable change in the mean value).

Fourth, **look beyond the FHP to improve spare part forecasts**. Further efforts to improve forecasting should focus on (admittedly harder) problems, such as the forecasting algorithms themselves, inventory policies (such as Peak Policy), and the information systems that contain them.

We understand that the Air Force already sought to improve its spare part forecasting system with the failed implementation of Expeditionary Combat Support System, and the Air Force is again investigating information system solutions to this (and other) issues. As new systems come online, one key question is when FHs should be used for forecasts. Empirical analyses can be performed to assess which items have a strong enough correlation to be useful, or, in other cases, where thresholds should be set such that FHP variance beyond a certain point would trigger some action. To the degree that these new systems provide insights into these questions and the ability to better calibrate spare part decisionmaking, the Air Force can realize some long-awaited benefits to readiness and cost-effectiveness.

Acknowledgments

Numerous people both within and outside the Air Force provided valuable assistance in support of our work. They are listed here with their ranks and positions as of the time of this research. We thank Lt Gen Bruce Litchfield, Commander, Air Force Sustainment Center (AFSC), for sponsoring this work. We also thank his staff for their time and support during this research, particularly Col John Kubinec of Vice Commander, Air Force Sustainment Center (AFSC/CV).

We thank our action officer, Ginger Hassen (at 448th Supply Chain Management Wing [SCMW]), for sharing knowledge about the Air Force's supply chain, for helping us find points of contact and data, and for quickly answering even the smallest questions.

We thank Brig Gen Allan Day, Deputy Director of Resource Integration and Logistics Chief Information Officer, Deputy Chief of Staff for Logistics, Engineering and Force Protection, Headquarters U.S. Air Force (AF/A4P) at Headquarters, Air Force for his support and interest.

We would especially like to thank Jennifer Matney at AFSC, who proved to be a veritable fountain of knowledge about the Air Force's financial system and working capital fund. We also thank Bob McCormick for providing helpful feedback on our report.

We thank Frank Washburn at 448 SCMW for making many staff members available, as well as Sandy Windsor, Wendy Walden, and their staffs and colleagues, also at 448 SCMW, for insightful discussions of surge policy and practice.

Finally, we thank RAND colleagues Michael Boito, Jerry Sollinger, and Regina Sandberg for their helpful edits.

Responsibility for the content of the document, analyses, and conclusions lies solely with the authors.

Abbreviations

ABCS	automated budget compilation system
AF	U.S. Air Force
AFMC	Air Force Materiel Command
AFSC	Air Force Sustainment Center
ANMIS	air navigation multiple indicators
CAF	Combat Air Forces
CER	Cost-Effective Readiness
DFA	demand forecast accuracy
DLR	depot-level reparable
DoD	U.S. Department of Defense
ELS	Enterprise Logistics Strategy
EXPRESS	Execution and Prioritization of Repair Support System
FH	flying hour
FHP	flying hour program
FY	fiscal year
HAF	Headquarters, Air Force
HQ	headquarters
HUD	head-up display
MAJCOM	major command
MDS	mission design series
MICAP	mission impaired capability awaiting parts
NIIN	National Item Identification Number
NRTS	not reparable this station
PAF	RAND Project AIR FORCE
POM	program objective memorandum
SME	subject-matter expert

SCMW   Supply Chain Management Wing

USAF   U.S. Air Force

Chapter One

Introduction

Between fiscal years (FYs) 1996 and 2011, U.S. Air Force (USAF) Non–Special Operations Forces fixed-wing operating and support spending increased at an average rate of 6.5 percent per year.[14] This steady rise in logistics spending crowds out other investments, such as modernization, unit readiness, and infrastructure. The increased spending, combined with decreasing budgets, prompted Air Force logistics leaders to urge the Air Force to adopt a new approach to balancing cost and effectiveness: Cost-Effective Readiness (CER)—which seeks to change the predominant mind-set from executing the budget to reducing the costs associated with producing readiness.

Delivering CER is a strategic priority of the Air Force's Enterprise Logistics Strategy (ELS).[15] One focus area of the ELS is to improve the accuracy of spare part supply requirements, i.e., demand forecasts. Documentation supporting the ELS identified a range of known and potential causes of forecast error,[16] including natural demand uncertainty, flying hour (FH) variance, phase-in or phase-out schedule issues, new items, and incorrect applications.

Many of the above factors fall within the scope and influence of Air Force Materiel Command (AFMC), in particular Air Force Sustainment Center (AFSC), which oversees all organic supply chain management and operations and operates all organic depot-level maintenance. One potential contributor to forecast error that falls outside AFMC's purview is what is known as FH program (FHP) variance. Each year, the Air Force estimates its FHP for pilot training and uses these estimates as a key input to forecast demand for depot-level reparable (DLR) spare parts. These FH forecasts drive downstream logistics decisions, in particular DLR spare buys and repair capacity. When the number of actual FHs (those flown during the year of execution) deviates from the number of planned FHs (those entered into the program objective memorandum [POM] process), this deviation is called FHP variance.

The FHP accounts for about $9 billion to $10 billion per year in spending, and the Air Force spends about $4 billion annually on spare parts (not all spare part spending is FHP-driven). Accurate demand forecasts ensure efficient and effective use of limited operating and support dollars and repair capacity. Spare part demand that is lower than forecast can result in excess inventory and cost, and demand that is higher than forecast can result in shortages that could

---

[14] Michael Boito, Thomas Light, Patrick Mills, and Laura H. Baldwin, *Managing U.S. Air Force Aircraft Operating and Support Costs: Insights from Recent RAND Analysis and Opportunities for the Future*, Santa Monica, Calif.: RAND Corporation, RR-1077-AF, 2016.

[15] U.S. Air Force, *2012–2022 Air Force Enterprise Logistics Strategy*, Version FY14.2, October 16, 2013b.

[16] Materials associated with CER use a range of terms to mean *error*: *variability*, *variance*, and *volatility*. In this report, we use the terms *variance* and *error* essentially interchangeably.

compromise readiness. AFSC analysts, in particular those in the 448th Supply Chain Management Wing (448 SCMW), more or less continually assess forecast accuracy and the root causes of forecast error.

In FY 2014, the AFSC asked RAND Project AIR FORCE (PAF) to assess the potential effect of FHP variance on cost and readiness. This analysis targets FHP variance as one driver of supply chain variance and seeks to do three things:

1. Identify the root causes of observed variance.
2. Quantify the effects of variance.
3. Suggest policy options to address variance and its effects.

The next chapter provides background on the FHP and demand forecasting. Chapter Three explains our approach to analyzing the downstream effects of FHP variance and offers our findings. Chapter Four provides our conclusions and recommendations, including some that offer greater benefit than correcting FHP variance.

Chapter Two

Background

This chapter provides some additional details on the FHP and describes how demand is forecast and what can affect that forecast. Those familiar with the program and processes might wish to skip the sections that deal with those topics. The third section of the chapter shows the variance that occurs in the FHP. The last section of the chapter describes how the Air Force forecasts demand for spare parts.

The Air Force Flying Hour Program

The Air Force FHP comprises the number of hours needed to attain and maintain combat readiness and capability for its aircrews, to test weapon systems and tactics, and to meet collateral requirements, such as air shows, demonstration rides for important personnel, and ferrying aircraft. Operations and training personnel at the major commands (MAJCOMs) determine the number of hours required annually.[17]

The Air Force estimates the budget for this program using a proportional cost model with two primary inputs: the number of FHs to be flown and a cost-per-FH factor. The product of the two inputs results in the expected FHP costs and estimated budget associated with each Air Force MDS.[18]

In the 1990s, in efforts to achieve greater accuracy in FH estimates, each MAJCOM switched to standardized methodologies that reflected the mission of that MAJCOM. The new models calculate FHs based on the number of pilots required to be combat mission-ready, basic mission-capable, or current with their training. The FH models also account for pilot experience, guidelines for mission types and weapon qualifications, special capability sorties, and collateral sorties.[19] Recent CER efforts have further refined these estimates.

---

[17] The Joint Mission Essential Task List, the Air Force task lists, and mission design series (MDS)–specific volumes of the Air Force Instruction 11-2 series are the foundational requirements that link aircrew training to tasks required to support combatant commanders. See U.S. Air Force, *Flying Hour Program Management*, Air Force Instruction 11-102, August 30, 2011.

[18] Tyler Hess, *Cost Forecasting Models for the Air Force Flying Hour Program*, Wright-Patterson Air Force Base, Ohio: Air Force Institute of Technology, March 2009.

[19] U.S. General Accounting Office, *Observations on the Air Force Flying Hour Program*, Washington, D.C., NSIAD-99-165, July 8, 1999.

The Forecasting and Demand Process

The FHP and demand forecasting both happen within a much larger context of strategic planning and budgeting. Figure 2.1 shows a simplified schematic of that larger context. In Figure 2.1, this process starts with strategic guidance. The President and the U.S. Department of Defense (DoD) assess the geopolitical and security environments and issue various guidance documents that contain national objectives and priorities. Senior Air Force leaders then translate those objectives, priorities, and scenarios into Air Force priorities and decisions, operationalized in guidance documents and budget decisions.[20] As part of the annual planning, programming, budgeting, and execution system process, also referred to as the POM process, the air staff, with inputs from the MAJCOMs, determines the annual FHP.

It is important to understand that the FHP, or at least the money for it, is ultimately a part of the POM, and it is subject to the same forces and constraints that affect the rest of the POM. The POM process can be unpredictable, and, when a cut is levied or funding must be found for a new requirement, it must come from somewhere. The FHP constitutes the largest single share of the Air Force's operations and maintenance appropriation (about 20 percent), and it is sometimes a

Figure 2.1. Strategic Context of the Flying Hour Program and Demand Forecasts

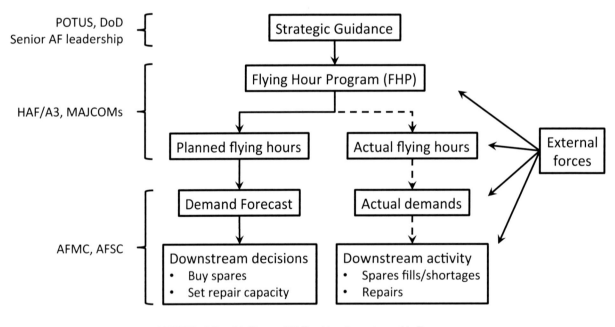

NOTES: AF = Air Force, HAF = Headquarters, Air Force.

---

[20] For more on the Air Force's strategic planning system, see U.S. Air Force, *Strategic Planning System*, Air Force Policy Directive 90-11, March 26, 2009.

target for short-term cuts when offsets must be found.[21] Sometimes, those cuts are subsequently restored in the year of execution as funding continues to shift, thus broadcasting a flying plan with known deficiencies to the logistics community, only to revise it once all the downstream decisions have been set in motion.

The CER working group highlighted a number of instances of FHP changes that created significant volatility:[22] in FY 2008, a 10-percent FH cut; in FY 2009, the Combat Air Forces (CAF) aircraft reduction; in FY 2011, the FHP optimized and reduced because of overseas contingency operations; in FY 2012, a 5-percent CAF FHP efficiency; and in FY 2013, the budget sequester.[23] Most of these were changes to the POM that occurred after initial FHP projections had been made, where fiscal or other external realities impinged on the FHP authors' original plans. And in nearly every case, FHs were *reduced* after the initial program was set, to make room for other priorities.

We now draw the reader's attention to the split below the FHP box in Figure 2.1. The left set of arrows and boxes represent the annual *plan* and the decisions that flow from it. Planned FHs drive the demand forecast, which, in turn, drives decisions to buy spare parts and set depot repair capacity (essentially organic depot labor levels) in the form of annual contracts with unionized government civilian personnel. At some point, because of lead times, those decisions cannot be reversed without additional investment of time, trouble, money, or some combination thereof. This side of the diagram comprises a significant aspect of the experience of logisticians and financial managers within the AFSC.

The right series of arrows and boxes represent the *actual* demands and the needs and activities that flow from them. The FHP recorded in the budget influences what is actually flown. But many other things can influence what is actually flown in a given time period or over the entire year—e.g., fiscal changes; contingency operations; and safety issues, such as the grounding of a fleet. The box on the right side labeled "External forces" depicts this influence. The operators setting the FHP in the POM process can conduct their process rigorously, but their product remains subject to many other forces. Thus, the line from the FHP to actual FHs is dashed. The line from actual FHs to actual demands (i.e., spare parts removed from an aircraft for further diagnosis and repair) is also dashed, which conveys the stochastic relationship between flying activity and spare part breaks (illuminated further below).

---

[21] For example, the FY 2013 Budget Control Act resulted in the Air Force grounding one-third of its combat fighter squadrons for three months. See David L. Goldfein, "Department of the Air Force Presentation to the Subcommittee on Readiness, United States House of Representatives Committee on Armed Services," February 12, 2016. Infrastructure sustainment restoration and modernization funding is another target, at about $3 billion to $4 billion per year.

[22] U.S. Air Force, *Cost Effective Readiness LOE #2: Flying Hour Program Inputs*, briefing, 2013a.

[23] The June 2011 Budget Control Act (also known as the "sequestration" or "budget sequester") called for sequestration of $1.2 trillion from the discretionary accounts of the federal budget beginning in January 2013 (which was later delayed to March 2013 by the American Taxpayer Relief Act of 2012).

Finally, the actual demands ultimately translate into actual spare part stock requests and repairs. This line is also dashed, to represent the probabilistic nature of that relationship. For example, a part that is removed might not be broken. A broken part might be repaired at unit level (thus not requiring depot repair). A part that is not repaired at unit level might have serviceable stock available. That part will, in any case, generate a demand for a depot repair. Then, at depot level, in addition to unit-level FH-driven demands, many other demands could materialize, including for depot-driven repairs, to fill stock levels, to supply other services, or to accommodate foreign military sales. A part coming in for repair might or might not be condemned. Many external factors disrupt the relationship between the FHP and actual downstream effects, including the actions and decisions of managers and maintainers to solve problems that are sometimes driven by inaccurate forecasts upstream.

Thus, because of the way the system is designed, the linkage between the FHP and downstream *decisions* (left side) is direct and straightforward, whereas the linkage between the FHP and downstream *activity* (right side) is more tenuous and is governed by a range of factors, each with inherent uncertainty. In other words, supply chain plans change whenever FH plans do, but the demands on the supply chain resulting from that flying might not.

Flying Hour Variance Data

Even with the process improvements implemented in the past, significant variations in FH execution occur. The data and plots in this section show this FHP variance from a number of perspectives.

Flying Hour Program Variance, by Aircraft Type

Figure 2.2 shows FH data for fixed-wing aircraft types for FY 2012. We have truncated the y-axis at 100 percent overflying, though two MDS exceeded this: The A-10 overflew by 55 times its original estimate; the UH-1H more than tripled its planned hours.[24]

---

[24] The Air Force has attempted for several years to retire the A-10 fleet and has made corresponding reductions to the A-10 FHP. Those attempts have thus far been met with political resistance, so the Air Force has had to put back FHs for the A-10 after the original FHP projections were made.

Figure 2.2. Flying Hour Variance Versus Number of Flying Hours

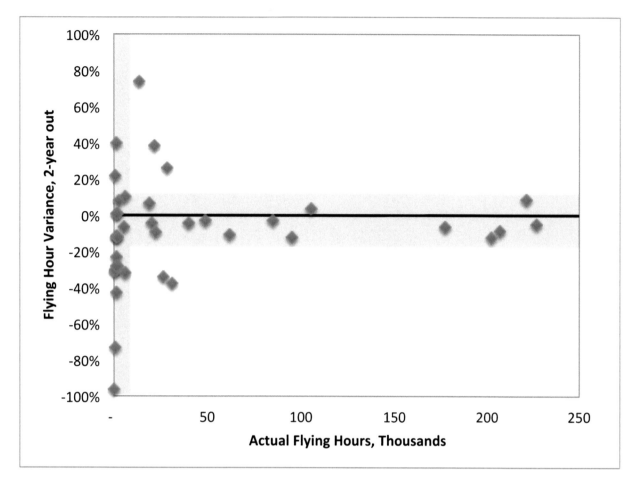

SOURCE: D200, Spares Requirements Review Board, 2012 (two-year-out forecast), provided by Headquarters (HQ) AFSC.
NOTES: Each diamond represents one MDS, showing the executed FHs (x-axis) compared with the FH variance as a percentage of forecast FHs, made two years out (y-axis). The farther away a diamond is from the centerline, the greater the variance from what was planned. A diamond above the line indicates that the MDS flew more hours than were budgeted, and below the line, fewer hours. The horizontal gray band indicates FHP variances between –12 percent and +10 percent; the vertical gray band indicates fewer than 6,000 FHs. We have truncated the y-axis at 100-percent overflying, although two MDS exceeded this: The A-10 overflew by 55 times its original estimate, and the UH-1H more than tripled its planned hours. D200 is a subsystem of the Air Force Requirements Management System. It is referred to hereafter in this document by its data system designator, D200.

One fairly obvious characteristic of the data is that, generally, the MDS with the fewest hours have the largest variances, and vice versa. In fact, we identify three nonoverlapping regions of this plot. Of the 40 MDS with forecast FHs in FY 2012, 22 had FHP variances between –12 percent and +10 percent (horizontal gray band), 13 were outside this range but had very few FHs (fewer than 6,000 hours, vertical gray band), and the remaining five MDS fit neither criterion (outside the gray bands).[25] So, then, most of the MDS had fairly low variances, very few FHs

---

[25] The C-21, E-8, B-1, and C-5 each had around 20,000 to 30,000 hours and error between either –20 and –40 percent or +20 and +40 percent.

(and thus ought to have a fairly small effect on the aggregate supply chain), or both. We looked at other years of data and found a similar pattern. Appendix B shows these data plots.

MDS in these three regions of the plot ought to have different types of effects. Very large programs (e.g., the KC-135, with 8-percent overflying in FY 2012) drive a large proportion of the supply chain activity and cost, and readiness effects of this fleet have far-reaching effects. Very small fleets might have little effect on the overall supply chain, but large perturbations in FHs could have significant program-unique implications. Across the population shown in Figure 2.2, more than 80 percent of the total FH error (i.e., aggregate hours of error across all aircraft types) was driven by MDS in the lower error range. This pattern is consistent for the other years of data.

Part-Level Flying Hour Program Variance

Figure 2.2 shows the weapon-system perspective, representing the idea that a single program makes a plan and flies its hours. For the supply chain, however, the important unit of account is parts. Many parts are installed on multiple MDS, so the total FH error experienced by many parts is actually the product of FHs on many different programs, which could be subject to many different forces.

Further, supply chain effects occur mostly at the part level. For part buys and stockage, only the exact part will do (setting aside suitable substitutes and the like). For repairs, part-level error can drive mismatches in consumable piece parts. But some depot labor is fungible because many personnel are cross-trained on multiple items, so some part-level error could be mitigated by this intrashop flexibility. For most of our analyses in Chapter Three, we perform our calculations at the part level, then simply aggregate effects up to higher levels.

Figure 2.3 shows part-level FH error data taken from D200 for FY 2012, for FH-driven DLRs. The display is identical to that in Figure 2.2, except that here, each diamond is a single part. Those parts with the most FHs generally had the least error, and vice versa.

This figure shows a similar pattern: The parts with most removals have FH error within a fairly tight band, and the greatest variance occurs in parts with the least removals. There are about 85 parts in this plot (out of about 4,500) that overflew by more than 100 percent. Only two MDS overflew by more than 100 percent (the A-10 and UH-1), so these outliers could be the result of data input problems, programmatic changes, or part phase-in or phase-out that did not align with expectations.

Figure 2.4 shows aggregate part-level FH error statistics from FYs 2009 through 2015.

Figure 2.3. Part-Level Flying Hour Error

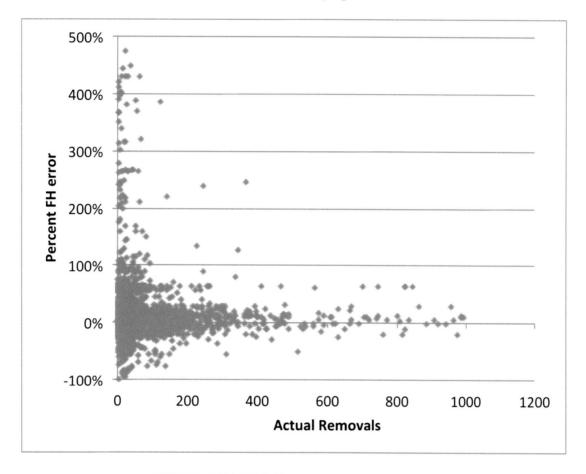

SOURCE: D200, FH DLRs only (two-year-out forecast).

Over this time period, except for FY 2014, over- and underflying errors were in roughly the same range (first and second columns for each year), about 5 to 15 percent. There is a slight upward trend in the absolute error. The obvious outlier here is FY 2014. Overflying skyrocketed, which dragged the net and absolute errors up as well. The culprit here was the budget sequester, which technically started in the middle of FY 2013. The significant underflying and net negative error for FY 2013 was largely the result of that: When the sequester hit, the FHP had long been set, and in fact half the fiscal year had already passed. At that point, the Air Force put the brakes on the FHP, fewer hours were flown, and thus there was a net underflying of the FHP.

In FY 2014, lower FH targets had already been set to better align with the sequester. But funding was eventually made available during the year, and some fleets tried to make up training deficits, thus overflying what were originally conservative FH goals.

Figure 2.4. Aggregate Flying Hour Error Statistics over Time

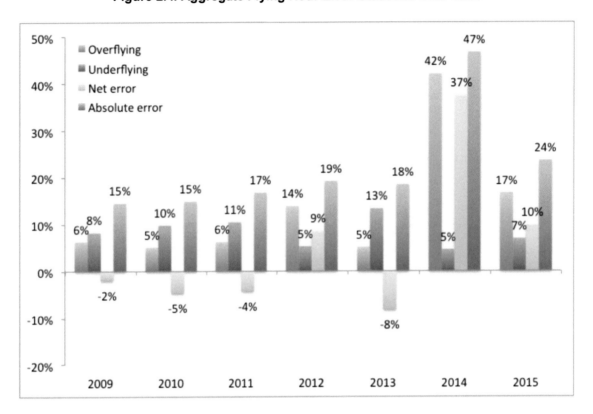

SOURCE: D200, FH DLRs only (two-years-out forecast).
NOTE: The blue columns show volume-weighted overflying error (i.e., overflown hours divided by total FHs).
The red columns show the same calculation but for underflying parts, the green columns the net error divided
by total FHs, and the purple columns the volume-weighted sum of the absolute values of the errors, divided
by total FHs (blue plus red columns). (Some numbers might not sum exactly because of rounding errors.)

Figure 2.5 shows these same part-level data from a slightly different perspective. The three
lines show the proportion of parts with various levels of *absolute* FH error (related to the last
column for each year in Figure 2.4). The huge spike in the 50-percent-or-greater line in FY 2014
mirrors the jump seen in Figure 2.4.

Figure 2.6 homes in on the two years of biggest impact to FH error (thus far) from the
sequester. For the FY 2013 line, the curve bends outward to the left, showing significantly more
underflying than in other years. The extent of underflying is not large: The highest three points
on the blue line are at –10 percent, 0, and –20 percent, respectively. But underflying still stands
out here against all other years, and especially FY 2014. We see the same underflying spike in
Figure 2.4, most clearly shown by the negative net error column, which stands out among other
recent years. Then, in Figure 2.5, FY 2014 has two distinct features. First, it has the least
underflying of all years; FY 2013 experienced the most underflying. Second, FY 2014 has a big
bump at 50-percent overflying. Overflying up to 40 percent does not appear high, but both
50 percent and 60 percent are higher than in other years. It appears that this 50- to 60-percent
overflying bubble is what drove the jump in errors we see in Figures 2.4 and 2.5.

Figure 2.5. Aggregate Absolute Flying Hour Error Statistics over Time

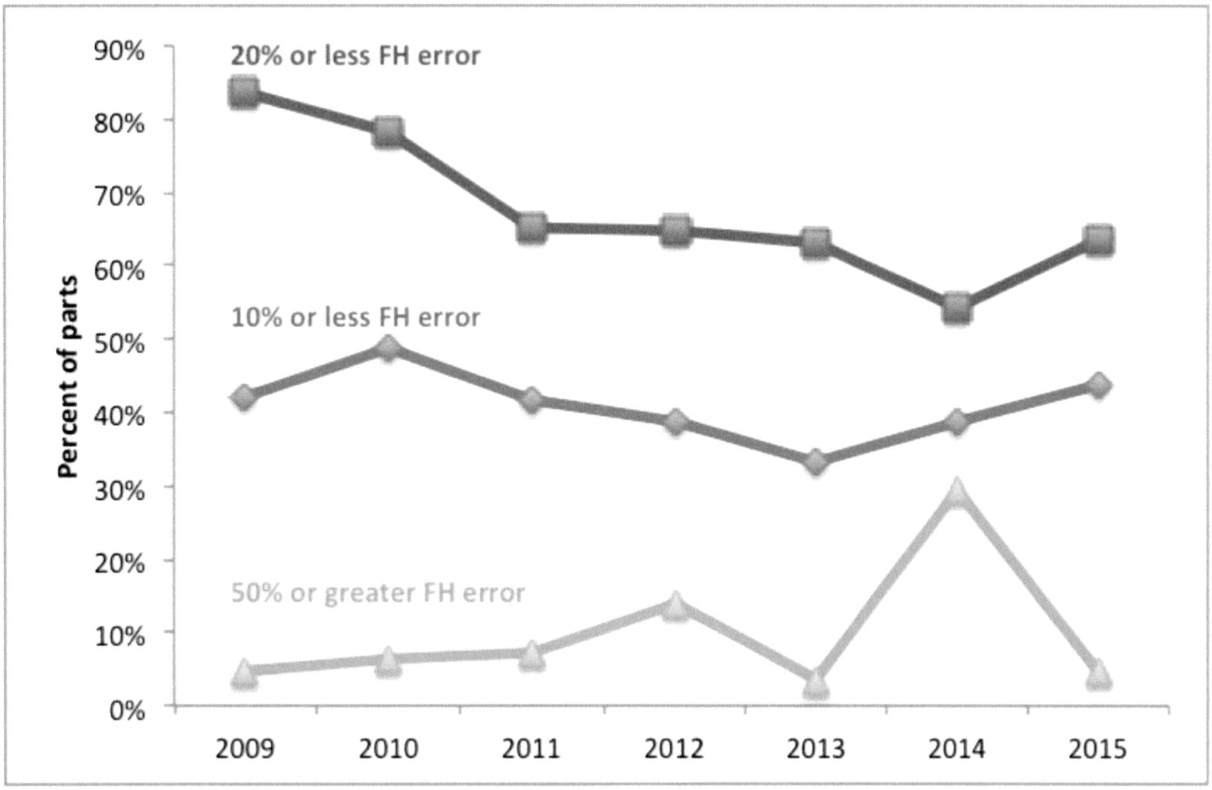

SOURCE: D200 data.
NOTE: The blue line with diamonds shows the proportion of all DLRs in the data set with 10-percent-or-less FH error. These are the most accurately predicted. The red line shows the proportion of DLRs in the data set with 20-percent-or-less FH error. We see that in FY 2009, 84 percent of DLRs had 20-percent or-less-FH error, but that steadily decreased to a low of 54 percent in FY 2014. The green line with triangles shows the proportion of DLRs with 50-percent-or-*greater* FH error. These are the least accurately predicted.

Thus, in FY 2013, many fleets put on the brakes late in the year to meet the sequester funding ceilings. And in FY 2014, after FH targets were reduced, some *limited portion* of the Air Force fleets overflew their sequester-driven targets to make up for training deficits.

Although disaggregated part-level data are appropriate for analysis, these aggregate metrics help give a general impression of what is happening. One can compare across years, populations, and other characteristics to see and diagnose more generally. But these aggregate statistics can mask the underlying character of the error, as is shown in Figures 2.2 and 2.3.

Figure 2.6. Frequency of Flying Hour Error Across Years

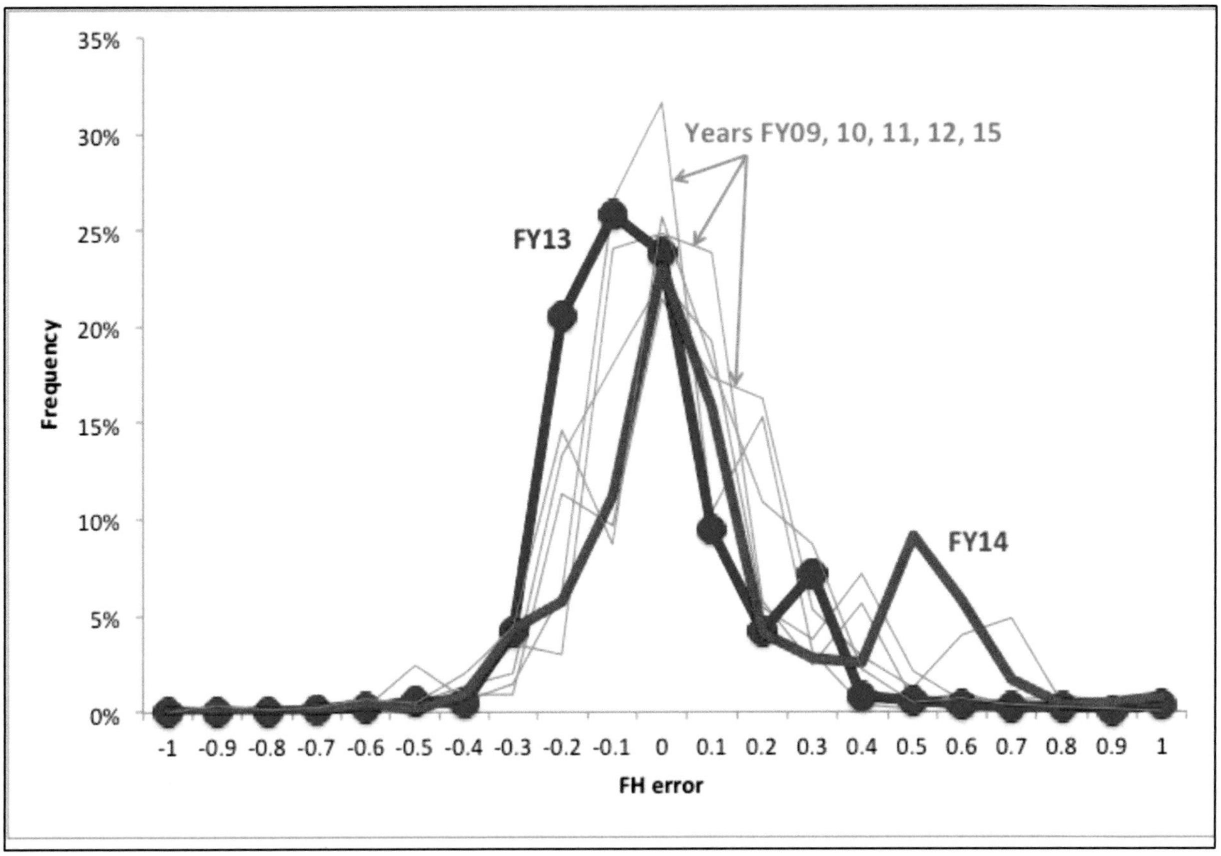

SOURCE: D200 data.
NOTE: Each line represents a single year of data. The x-axis shows the percentage FH error from –100 percent to +100 percent—although overflying of greater than 100 percent is possible, we truncated the display for legibility. The thin gray lines show data for FYs 2009–2012 and 2015. The thick blue line with circles shows FY 2013; the thick red line shows FY 2014.

Summary of Flying Hour Variance Findings

In light of the data analysis presented above and our conversations with Air Force subject-matter experts (SMEs) about the dynamics of FHP decisionmaking, we group the range of causes of FHP variance into three broad categories.

The first category, simple planning error, is a natural part of any process. Even without the external forces described above, forecasting the FHP exactly is impossible, given uncertainties in the number of pilots to train, the number of sorties they will need to or be able to fly, the maintenance status of their aircraft, and so on. We observed above that the majority of total FH error was driven by large numbers of parts with relatively small individual errors, apparently the result of natural uncertainty inherent in the planning process.

External causes come from outside the service: Contingency operations (or changes to ongoing operations) and congressional action are the two most dominant in recent history.

Sequestration appears to be the main example of congressional action that registered in enterprise-level metrics, but contingency operations have caused disruptions for several years.

The third category is internal Air Force decisions. These are decisions made by Air Force leaders regarding force structure, the budget as a whole, or the FHP specifically but that are made after the original FH forecast. Examples of this are the wide-scale FH cuts in FYs 2008, 2011, and 2012, and the CAF aircraft reduction in FY 2009. These are some of the most disruptive examples identified in the original CER effort and are sometimes large enough to register in enterprise-wide error metrics. The anticipated retirement of the A-10 could have been in this third category (and might not have been too disruptive to the FHP) had Congress not intervened. The reversal of the retirement decision (which puts this example in the second category) and uncertainty in A-10 force structure repeatedly led to low FH forecasts and severe overflying.

In a later section, we discuss how policy solutions and expected outcomes might be shaped by the nature of these different causes.

Thus far, we have discussed error only in FH forecasts. Now, we turn our attention to forecasting removals. As described above, FH error is but one contributor to error in forecasting removals.

How the Air Force Forecasts Spare Part Demand

The Air Force's logistics system depends on a reliable supply of spare parts, and forecasting demands for those spare parts has a long history. Since the earliest quantitative examinations of demand patterns for aircraft spare parts in the Air Force, it has been observed that the supply chain is challenged by a demand for parts that is inherently uncertain.[26] Part of this uncertainty arises from the stochastic nature of peacetime demands, and part arises from external fluctuations in demand such as those induced by contingency operations.[27] To address these sources of uncertainty and thereby reduce supply chain risk, there is a long history of two branches of emphasis for mitigation. One emphasis is on improved forecasting of demand. Efforts to enhance the accuracy of forecasts include sophisticated models,[28] extensive planning for driving factors

---

[26] The earliest RAND paper on the topic is Bernice B. Brown and Murray A. Geisler, *Analysis of the Demand Patterns for B-47 Airframe Parts at Air Base Level*, Santa Monica, Calif.: RAND Corporation, RM-1297, 1954. See also Bernice B. Brown, *Characteristics of Demand for Aircraft Spare Parts*, Santa Monica, Calif.: RAND Corporation, R-292, 1956.

[27] Called "statistical uncertainty" and "state-of-the-world uncertainty"; see James S. Hodges and Raymond A. Pyles, *Onward Through the Fog: Uncertainty and Management Adaptation in Systems Analysis and Design*, Santa Monica, Calif.: RAND Corporation, R-3760-AF/A/OSD, 1990.

[28] See, e.g., Craig C. Sherbrooke, *METRIC: A Multi-Echelon Technique for Recoverable Item Control*, Santa Monica, Calif: RAND Corporation, RM-5078-PR, 1966; Craig C. Sherbrooke, "METRIC: A Multi-Echelon Technique for Recoverable Item Control," *Operations Research*, Vol. 16, No. 1, 1968, pp. 122–141; Raymond A. Pyles, *The Dyna-METRIC Readiness Assessment Model: Motivation, Capabilities, and Use*, Santa Monica, Calif.: RAND Corporation, R-2886-AF, 1984; John L. Adams, John B. Abell, and Karen E. Isaacson, *Modeling and Forecasting the Demand for Aircraft Recoverable Spare Parts*, Santa Monica, Calif.: RAND Corporation, R-4211-AF/OSD, 1993; John B. Abell, *Estimating Requirements for Aircraft Recoverable Spares and Depot Repair:*

such as flying hour programs, and increased data collection and assessment. The other emphasis is on adaptive, robust processes that provide sufficient resiliency in the supply chain to absorb inevitable uncertainties, especially for parts with low demand.[29] After years of theoretical research and real-world experience, the consensus view is to predict as best as possible and maintain enough resiliency in the supply chain to accommodate the rest.[30] This deep history of research into spare part demand reveals that those demands often do not adhere to the assumptions of the forecasting models used to predict them,[31] but forecasting must be done.

For parts that the Air Force deems to be driven primarily by flying (i.e., the key cause of failure is flying activity, rather than a change to an item based on a schedule), it uses a model in which the number of historical FHs for a part is divided by the number of historical removals for that part to arrive at what is referred to as the *removal rate*. To produce such a forecast, the forecasting model uses the eight most recent quarters of data for FHs and removals to forecast demands for future quarters.[32] This relationship is described by the following:[33]

$$
\begin{aligned}
&number\ of\ failures\ in\ future\ quarter \\
&= projected\ flying\ hours\ \times\ \frac{number\ of\ failures\ in\ the\ past\ two\ years}{flying\ hours\ in\ the\ past\ two\ years}.
\end{aligned}
$$

The number of failures (or removals) per FH (the term on the right side) is often called the *removal rate*. This model assumes a perfectly linear relationship between FHs and removals, with no other explanatory variables.

Figure 2.7 shows D200 data with actual FHs and removals for two example items (by National Item Identification Number [NIIN]). The left panel shows NIIN 001249409 from the F-15 head-up displays (HUDs) and air navigation multiple indicators (ANMIS) shop; the right panel NIIN 001095725 from the KC-135 boom shop. Each diamond represents a single quarter of data for all items within that shop. Each plot displays 41 quarters of data, shown with blue

Executive Summary, Santa Monica, Calif.: RAND Corporation, R-4215-AF, 1993; F. Michael Slay, Tovey C. Bachman, Robert C. Kline, T. J. O'Malley, Frank L. Eichorn, and Randall M. King, *Optimizing Spares Support: The Aircraft Sustainability Model*, McLean, Va.: Logistics Management Institute, AF501MR1, 1996; and John A. Muckstadt, *Analysis and Algorithms for Service Parts Supply Chains*, New York: Springer, 2005.

[29] See, for example, Gordon B. Crawford, *Variability in the Demands for Aircraft Spare Parts: Its Magnitude and Implications*, Santa Monica, Calif.: RAND Corporation, R-3318-AF, 1988; Hodges and Pyles (1990); Irv K. Cohen, John B. Abell, and Thomas F. Lippiatt, *Coupling Logistics to Operations to Meet Uncertainty and the Threat (CLOUT): An Overview*, Santa Monica, Calif.: RAND Corporation, R-3979-AF, 1991.

[30] See especially Hodges and Pyles (1990).

[31] See Crawford (1988).

[32] These calculations for reparable spares requirements are computed in D200A.

[33] The eight-quarter moving average is one of multiple options within D200. D200 also has the ability to compute requirements using a four-quarter moving average, exponential smoothing, and a predictive logistics model that utilizes a regression technique. We appreciate RAND colleagues Don Snyder, Sarah Nowak, and Kristin F. Lynch for sharing their unpublished research with us.

diamonds, spanning FYs 2002–2012. The red squares show the eight most recent quarters of data, which would be used in a D200 forecast.

Figure 2.7. Example Part with Flying Hours and Removals

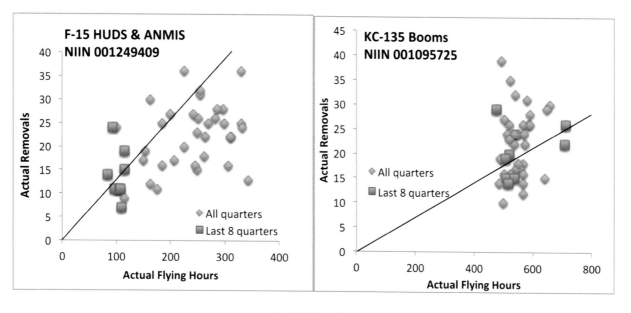

The lines on each plot show the computed removal rate using the equation above and the red data points. One can see how much variance occurs between this computed relationship and the relationship in any given quarter. For the F-15 part, the computed removal rate using the past eight quarters of data would be 13.6 per 100 FHs, with a range of 6.4 to 25.8 removals per 100 FHs over that time period. For the KC-135 part, the computed removal rate using the past eight quarters of data would be 3.56 per 100 FHs, with a range of 2.7 to 6.1 removals per 100 FHs over that time period.

For both eight-quarter data sets, there is no discernible relationship between FHs and removals. The forecast relationship bears literally no relationship to the data. In the left panel, there is some correlation if all data points are included. But in that case, the y-intercept is nonzero. Thus, the D200 relationship greatly overstates the effect of FHs alone on removals. Most NIINs look something like these two plots.

The part-level relationship is the only one appropriate for informing part buys. This suggests that D200's presumed relationship between FHs and removals is simply not very useful in many, if not most, cases.

The relationship between FHs and removals sometimes improves at higher levels of aggregation. For example, the shop for the part on the left of Figure 2.7 shows fairly strong correlation with an easily visible pattern, whereas the shop for the part on the right does not.

15

Because of the flexibility of labor, shop-level correlation suggests that FHs could be a useful input for setting shop capacity in some cases.

However, we highlight the fact that many more factors contribute to error in the chain of events between the number of removals and the number of parts actually arriving for repair at a depot. These include not-reparable-this-station (NRTS) rates,[34] condemnation rates, Execution and Prioritization of Repair Support System (EXPRESS) prioritization, depot-generated demands, and other sources of demand (e.g., other service, foreign military sales).

Consider the effect of this tenuous or nonexistent part-level relationship between FHs and removals on the overall chain of events in Figure 2.1. The left side of the diagram depicts a direct relationship between FHs and downstream actions, but we see now that this relationship, at the disaggregated level at which it matters, is virtually nonexistent.

With that backdrop, we now shift our attention to assessing the accuracy of these parts forecasts.

How the Air Force Currently Assesses Forecast Accuracy

The Air Force (through the 448 SCMW) assesses its forecast accuracy to identify and analyze problems, track progress toward goals, and report performance to HQ Air Force and DoD. It does so with several metrics. Demand forecast accuracy (DFA) assesses the total volume of error in parts removals. There are two versions of this metric: one called *demand volume–weighted DFA* and one called *dollar demand–weighted DFA*.

The equation for the demand volume–weighted DFA is shown here:

$$DFA\ (Vol.) = 1 - \frac{\sum_{NIIN} |actual\ demand - predicted\ demand|}{\sum_{NIIN} actual\ demand}.$$

Note the absolute value in the numerator. This means that demand volume–weighted DFA takes the total error, not the net error. Note also that the numerator and denominator are summed separately. This means that the metric is volume-weighted, so a part with 1,000 demands per year counts toward the metric more than a part with ten demands. This emphasizes supply chain effects, not merely forecast process accuracy.

The dollar demand–weighted equation simply multiplies the demands by the dollar value of the part:

$$DFA\ (Dol.) = 1 - \frac{\sum_{NIIN} (|actual\ demand - predicted\ demand| \times value)}{\sum_{NIIN} (actual\ demand \times value)}.$$

---

[34] "Not reparable this station" is a category that is applied to an inoperable part when repair of the part is not authorized at the installation or cannot be accomplished because authorized equipment, tools, facilities, or qualified personnel are not available.

This calculation increases the influence of high-dollar parts. This metric is DoD-mandated and is the default metric used by the Air Force. Its purpose, in the words of the U.S. Government Accountability Office, is to assist "in better understanding effects on business outcomes."[35]

The third metric is bias. Below is the equation used to calculate bias:[36]

$$Bias\ (Dol.) = \frac{\sum_{NIIN} [predicted\ demand - actual\ demand\ ) \times value]}{\sum_{NIIN} (actual\ demand \times value)}.$$

The main difference between the second and third equations is the absolute value in the numerator.[37] Here, the positive and negative errors cancel each other out instead of adding up.

Figure 2.8 shows data for these three metrics for all USAF-managed spare parts (not just those driven by FHs) from FYs 2008 to 2014 (calculated on a semiannual basis).[38] The dotted blue line shows the demand volume–weighted value, the solid red line shows the dollar demand–weighted value, and the black line with circles the net bias. The demand volume–weighted DFA increased from about 30 percent in FY 2008 to about 50 percent in FY 2012, when it appeared to level off.

Personnel from 448 SCMW informed us that the wing scrubbed the forecast data during the period of DFA increase and, in large part, reversed decisions made by individual managers to adjust computer-generated forecast values.[39] Many of these management decisions were hedges that increased forecasts above automated ones: These changes generally resulted in higher (and thus more robust) resource levels but also increased error by overplanning resources relative to actual demands.[40] The scrubbing of these planning factors brought forecasts more in line with automated ones and increased their accuracy.

The dollar demand–weighted DFA remained more stable over this time period. Given that the Air Force, until recently, used primarily the demand volume–weighted metric, it makes sense that improvements were seen in that metric. Efforts focused on improving that metric would favor high-demand items, which usually have lower costs.

---

[35] U.S. Government Accountability Office, *Defense Inventory: Actions Needed to Improve the Defense Logistics Agency's Inventory Management*, Washington, D.C., GAO-14-495, June 2014.

[36] Email communication with 448 SCMW personnel on October 30, 2015.

[37] The other difference is that the bias equation does not have a "one minus" in the numerator.

[38] Values taken from PowerPoint briefings provided by 420th Supply Chain Management Squadron personnel on October 7, 2014, and February 20, 2015. Josh Moore, *448th Supply Chain Management Wing Demand Forecast Accuracy Sep 08–Mar 14 Briefing*, October 2014, and Josh Moore, *448th Supply Chain Management Wing FY14 DFA Results Briefing*, January 14, 2015.

[39] Telephone interview with 448 SCMW personnel on February 20, 2015.

[40] PAF analysis of D200 data for DLRs driven by FHs shows a systematic overplanning bias in removal rates. This bias was as high as 40 percent in FY 2008 and gradually declined to about 5 percent in FY 2012.

Figure 2.8. Depot-Level Reparable Demand Forecast Accuracy Metrics

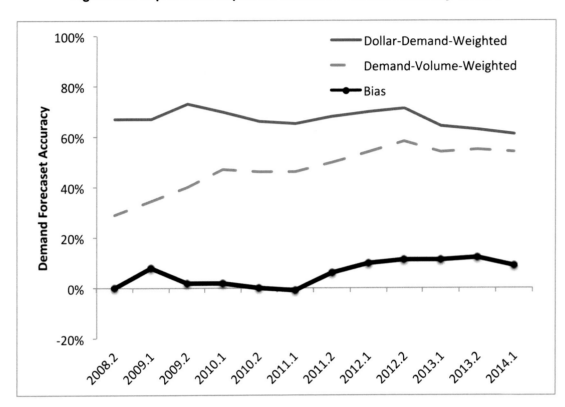

SOURCE: 448 SCMW, all reparable spare parts.

In FY 2014, the Air Force transitioned to a standard dollar demand–weighted metric developed by the Office of the Secretary of Defense. Presumably, improvement efforts will shift to those higher-dollar items that drive that metric.

In Chapter Three, the next question we ask is how much of the error we observe in the demand forecasts is actually driven by the error in the number of FHs we observe.

Quantifying the Negative Effects of Flying Hour Program Variance

In this chapter, we investigate the effect of FHP variance on demand forecast accuracy (DFA), opportunity cost in the budgeting process, long-term financial costs, and readiness.

Relationship Between Flying Hour Program Variance and Forecast Accuracy

Figure 2.1 and the surrounding text explain that the FHP drives the spare part forecast but also note that a host of other factors influence actual spare part demand. The project team modeled the extent to which departures from the planned FH program affect the accuracy of forecasts for flying-related spare parts.

How We Modeled the Effect of Flying Hour Program Variance on Forecast Accuracy

Our objective was to estimate DFA as a function of different levels of over- or underflying. To do this, for each item in our data set (described in data sources in this chapter), we needed to estimate predicted and actual demands for various levels of over- and underflying. Data from D200 tell us actual demands, actual FHs, and the demands per FH rate that were historically used to generate predicted demands. We did not want to assume any particular model for the relationship between *actual FHs* and *actual demands* because considerable uncertainty and, accordingly, lack of consensus surround this relationship. However, the relationship between *predicted demands* and *predicted FHs* is straightforward in Air Force models.

For FH-driven items, the Air Force assumes that (predicted demands) = (predicted FHs) * (break rate), described in Chapter Two. This formulation enabled us to examine counterfactual situations where predicted FHs differed from actual FHs by different percentages.

We used D200 data from FYs 2013 through 2015 to find the following three terms:

- projected break rates per FH
- actual item demands for FH-driven DLRs (total annual demands for each year)
- actual FHs for each item.

We used break rates estimated four quarters before actual demands occurred, representing a prediction one year out. For various levels of over- or underflying, we calculated what predicted item demands *would have been*. This calculation involved the following steps.

We assumed that if the Air Force had overflown x percent, that means

$$(actual\ FH) = \left(1 + \frac{x}{100}\right) \times (predicted\ FH) .$$

Therefore,

$$(predicted\ FH) = \frac{(actual\ FH)}{\left(1 + \frac{x}{100}\right)}.$$

Similarly, underflying by *x* percent means that

$$(predicted\ FH) = \frac{(actual\ FH)}{\left(1 - \frac{x}{100}\right)}.$$

Using D200 formulas, we then calculated for different levels of over- and underflying:

$$(predicted\ demands) = (break\ rate) \times (predicted\ FH)$$

As a simple example, suppose that in D200, an item had 120 demands in FY 2015, flew ten hours, and in FY 2014 had an estimated demand rate of ten demands per FH. We would estimate various levels of error in the predicted demand for that item as follows:

- If FHs had been predicted perfectly in FY 2014, the predicted number of demands for FY 2015 would have been 10 * 10 = 100, and the absolute error in the demand for that item would have been 20.
- If the item had *overflown* by 10 percent in FY 2015, the predicted number of FHs in FY 2014 would have been 10/(1.1) = 9.1. If the predicted number of FHs were 9.1, predicted demand would have been 91, and the absolute error would have been 29.
- If the item had *underflown* by 10 percent in FY 2015, predicted FHs would have been 10/0.9 = 11.1 in FY 2014, the Air Force would have predicted 111 demands for the item, and the absolute error for that item would have been 9.

Results

Figure 3.1 shows parts forecast (y-axis) as a function of FH error (x-axis). The green columns to the left or right of the middle show the DFA for different levels of under- and overflying, respectively. The middle green column shows that if FHs were forecast perfectly (i.e., with zero error), the DFA for FH DLRs would be expected to be 62 percent. That value is simply the demand volume–weighted DFA for parts with zero FHP variance.

At high levels of FHP variance, the DFA significantly diminishes. However, the DFA changes little for relatively small errors in predicted FHs (–10 percent to +10 percent), which is consistent with recent experience.

Figure 3.1. Relationship Between Flying Hour Program Variance and Demand Forecast Error

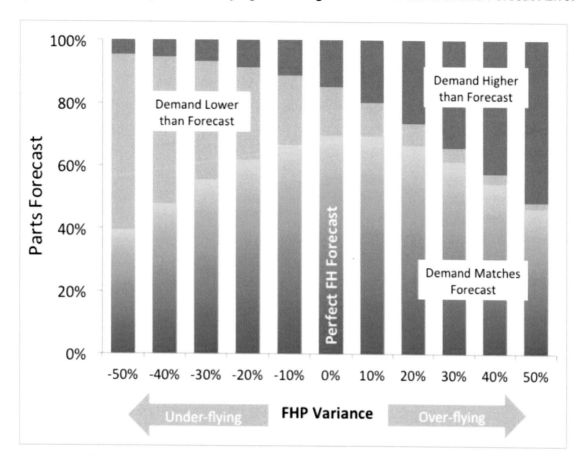

SOURCE: D200 data for FH DLRs, FYs 2013–2015 (two-year-out forecast).

As one moves to the left or right, forecast error increases as FH error does. But even at 50-percent flying error, other sources of error still constitute the majority of total forecast error. If all parts were underflown by 50 percent (leftmost column), total error would be 60 percent (sum of the blue and purple column sections), only 30 points of which are attributable to FHP variance.[41] If all parts were overflown by 50 percent (rightmost column), total error would be 53 percent, only 23 points of which are attributable to FHP variance.

The blue sections show forecast errors for which actual demand was lower than forecast DLR demand: These errors result in holding parts inventory in excess of near-term needs, increasing cost. The purple sections show errors for which actual DLR removals outstrip plans: Here, DLR shortages might compromise readiness. At zero FHP variance, these two errors are about equal.

---

[41] In this case, the error driven by FHs is slightly less than 30 percent, and the other error is slightly more than 30 percent. The total is slightly more than 60 percent but rounds to 60 percent.

The inherent "noisiness" occurs because of the underlying fact that FHs simply do not correlate very strongly to removals, as shown in Figure 2.5.[42] This conclusion is reinforced by the balance of over- and underplanning error: As one moves to the left, underflying error increases and overflying error decreases, and vice versa. But even at 50-percent overplanning of FHs, a small percentage of DLRs remains underplanned, and vice versa.[43]

On the whole, forecast error induced by FHs added about five points to the baseline (i.e., non–FH induced) forecast error, which translates to an increase of about 15 percent.[44] As shown here, higher FHP variance (such as the extreme overflying during FY 2014) would have larger effects on DFA.

If FHs were known perfectly, would forecasts be improved? Until recently, no. Figure 3.2 compares the DFA for planned with actual FHs. The height of each bar shows the DFA; each pair compares the value for planned FHs with actual FHs. The left blue column shows the value for planned FHs, the right red column for actual FHs. The figure shows the comparisons for FY 2010–2015 flying activity, with forecasts one year out.

Figure 3.2. Demand Forecast Accuracy for Planned Versus Actual Number of Flying Hours

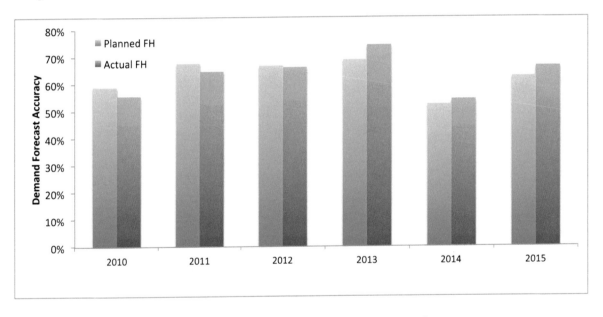

SOURCE: D200, FH DLRs (two-year-out forecast).

---

[42] This conclusion is corroborated by our own aggregate calculations using D200 data, as well as by analysis from Analysis Directorate, Strategic Plans, Programs, Requirements and Assessments Division, Air Force Materiel Command (AFMC/A9A).

[43] The reader might notice that the maximum DFA in Figure 3.1 is observed at both 0-percent and 10-percent overflying. We found systematic overplanning bias in the D200 data. So, on the whole, overflying by about 10 percent actually corrects for the overplanning error already in the system. This is discussed in relation with Figure 3.2.

[44] The baseline error (parts error when FH variance is 0) shown in Figure 3.1 is 30 percent. In earlier years (e.g., FYs 2010–2012), that baseline error was nearly 40 percent.

In the first three pairs in Figure 3.2, forecasting part removals using the actual FHs (i.e., with perfect knowledge of future flying) would have resulted in a *lower* DFA than with the original, planned FHs. The difference is pronounced for FY 2010 FHs and is barely visible by FY 2012. Then, for FYs 2013 through 2015, forecasting using actual FHs would have resulted in a higher DFA than with planned FHs, as intuition would suggest. We observed in the data that, in these earlier years, manual intervention was causing systematic overplanning to ensure sufficient inventory levels. That arguably improves readiness but at the expense of stocking more parts and of the DFA metric. We were told that the amount of this manual intervention had been reduced in recent years, thus reducing systematic overplanning bias and increasing overall DFA.[45]

Even though FHP variance has generally had modest effects on enterprise-level DFA (with the exception of the sequester in FY 2014), two concerns remain. The first is that even small enterprise-level perturbations could drive large cost or readiness issues. After all, the FHP drives many billions of dollars in spending every year. The other is that some individual programs (i.e., aircraft fleets) do experience large variance, and these drive real-world decisions that could later prove to be costly.

We now investigate other potential downstream effects of FHP variance: opportunity cost, financial cost, and readiness.

Opportunity Cost

Opportunity cost represents a lost opportunity in the annual programming process. For example, if a forecast with zero error resulted in logistics spending of $1 billion in a given year, but the actual forecast (with some error) resulted in logistics spending of $1.2 billion, we can say that there was an opportunity cost of $200 million. The USAF could have spent that money on something else: readiness for another weapon system, infrastructure repairs, or any number of other things. This is not to say that the "extra" $200 million was wasted. Indeed, if the money were spent on reparable spare parts, the USAF no doubt would use them, would get readiness value out of them, and could reuse them for some time. But budgeting the "right" amount provides enough money to support expected levels of readiness and takes nothing additional away from other competing priorities. The opportunity cost as described here simply represents a number of dollars the USAF *could have* spent differently. Said another way, that is money that the logistics community could "give back" to the USAF. In this section, we will also use the term *budget error* interchangeably with *opportunity cost* to stress that this is an artifact of the budget process, not necessarily a financial cost to the Air Force, and to help us distinguish between positive errors (i.e., spending too much) and negative errors (i.e., spending too little).

Because the opportunity cost is a figure in the annual programming process, we take the net over- and underspending resulting from over- and underplanning. If the USAF bought too many

---

[45] This shift in manual intervention was corroborated by 448 SCMW personnel.

of part A, spending $10 more than it should have, and too few of part B, spending $5 less than it should have, the opportunity cost is $5, not $10. With a perfectly accurate forecast, the total logistics spending would go down by only $5 (in this case). We are not yet concerned with other downstream effects of either over- or underplanning, which we discuss later.

To assess the potential opportunity costs of FHP variance, we developed a computer simulation to replicate DLR spares buy or repair decisions and quantify spending driven by planning error. (See Appendix A for technical details.) Our model allows us to simulate buy-or-repair decisions, and thus total logistics spending, based on different levels of FHP variance. For example, 10-percent overflying results in spending of about $270 million more on spare parts than if the FHP were flown exactly as planned. We applied the mathematical relationships between FHP variance and logistics spending to recent data on FHP variance. Figure 3.3 shows the results.

In Figure 3.3, the purple columns represent positive budget error, the spending in the budget for overforecast parts. Every time a part is overforecast relative to a perfect forecast, it drives resource allocation for repair capacity and part buys beyond what current analytic tools (for repair planning and part stockage) suggest is necessary to meet near-term needs (i.e., until the next funding cycle). With a perfect forecast, this is spending the Air Force could avoid (at least in that fiscal year) and put back on the table for other priorities. The green columns represent the counterpart of the purple column: the spending in the budget for underforecast parts (i.e., negative budget error). This represents a budget gap: Here the budget contains too little funding to support the expected demand for spare parts. The black dotted line represents the net budget error, or opportunity cost. As explained above, because these are simply budget dollars and not actual spare parts, the dollars are fungible.

In Figure 3.3, we see that the net opportunity cost was positive about half the time between FYs 2009 and 2015. The most extreme year of positive opportunity cost was FY 2013, the first year of the sequester. During these years of positive opportunity cost, there was net underflying (i.e., overforecasting) of FHs, meaning that *more* money was being programmed for buying and repairing spare parts on balance than was needed to support flying operations at expected levels of readiness. (Compare the pattern to Figure 2.4. Caution: Underflying equates to overbudgeting; the terminology can be confusing.) At the time this project was launched in FY 2014 by HQ AFSC, most of the center's recent history showed overbudgeting, creating a sense that money was consistently being left on the table in the budgeting process—in the case of FYs 2009–2011, about $100 million per year, a nontrivial amount.[46]

---

[46] HQ AFMC/A9A analysis in this same time range estimated the opportunity cost of spares (i.e., overbudgeting) of $60 million to $170 million per year for FY 2011–2012 examples.

Figure 3.3. Opportunity Cost of Flying Hour Program Variance

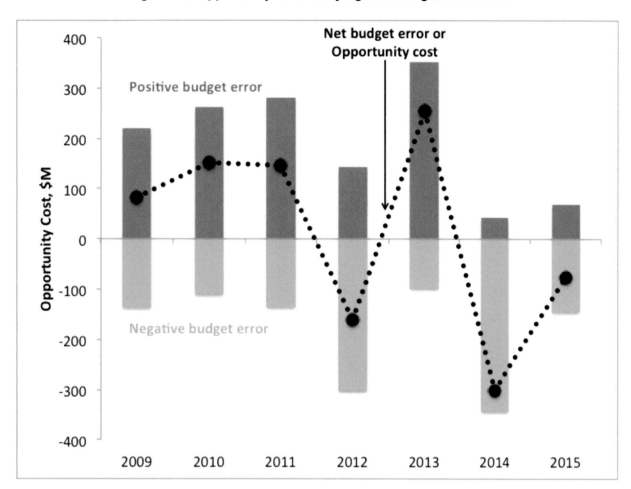

For FYs 2013–2015, the fluctuations mirror the patterns shown in Chapter Two's analysis of aggregate FH error. In FY 2013, FHs were drastically cut back. Normally, this would lead to overbudgeting, which Figure 3.3 is designed to reflect. In reality, the money was removed during the year of execution, so what this figure shows as about $250 million in net opportunity cost was probably much lower because the FHs were reduced to meet the artificially imposed budget ceiling. FY 2014 brought significant underbudgeting: Fewer hours, and thus dollars, were programmed to meet the ceiling, then some select fleets overflew (see the thick red line in Figure 2.6 for FY 2014).

There are two main reasons for such a large dollar effect from what appears to be a relatively small change in FH forecast error: (1) In recent years, total spending on DLR spares averages more than $3 billion per year, and (2) budgetary effects are determined by spare part forecasts, which are directly tied to FH forecasts and are not affected by any differences between planned and actual DLR demands. So overforecasting by a certain amount has a more or less proportional effect on overbudgeting, and vice versa. And even a small fraction of $3 billion is a significant amount of funding.

Financial Costs of Flying Hour Program Variance

In theory, both over- and underplanning can exact financial costs. In the context of the DLR spares supply chain, overplanning results in the purchase of more parts than are called for within the planning horizon, and these carry storage and obsolescence costs. Interviews with AFSC SMEs suggest that unused repair capacity (as a result of overplanning) is a lesser concern.[47] For underplanning, work-arounds can occur when the parts or labor needed to perform a repair are not available once a reparable part has been inducted into the depot.[48] However, FH underplanning error alone does not likely cause many work-arounds, because the relationship between FHs and removals is already weak, and many other sources of variance between part removal and a depot shop can result in a work-around. Thus, we focus our analysis of financial costs on parts costs that result from overplanning, using the same simulation referenced above.

It is useful to group DLR requirements into two categories. The first requirement is for parts that fail during service and must be replaced. The second is parts needed to achieve specified readiness positions ("holes" in weapon systems and establishing war reserve stocks) by a future time. We call these requirements, respectively, *keep up* and *catch up*. In general, unless there is a surplus of stock or reduced operational aircraft availability is desired, the keep-up requirements must be satisfied, whereas catch-up requirements are policy-driven variables.[49] In our simulation, we hold any catch-up holes constant and define the buy-or-repair requirements according to the keep-up requirement. Thus, we estimate unneeded inventory driven by FHs and other error according to the keep-up requirement.[50]

With no FH error, the simulation generates a one-year DLR buy of about $800 million. If FHs for all parts are overplanned by 10 percent, for example, the one-year buy increases by $250 million, on average. (Figure A.3 shows that these estimates have significant uncertainty because of uncertainty in demands.)

Of the additional $250 million from that notional 10-percent overplanning, our calculations show that $160 million would buy DLRs that would support current requirements (i.e., including

---

[47] The depot system has two mechanisms that mitigate overplanning. Initial workload estimates driven by customer-forecast demand always exceed historical levels of production (i.e., what actually "drives in"). Thus, in their annual planning process, called *requirements review and depot determination*, depot planners modulate initial workload estimates to align with historical production levels. Additionally, planners decrement labor levels by 5 percent of the stated requirement, expecting to make up any shortfalls using overtime. Further, when asked about cases in which maintainers really were "standing around," the only example that depot personnel could produce was an occasion when a removal rate planning factor was not updated, thus leaving repair capacity for a line that was expected to be shut down.

[48] Some examples of work-arounds are part cannibalizations, local manufacture of a part, and rerouting a repair to accomplish a portion of repair tasks until resources are available to complete it.

[49] Richard Hillestad, Robert Kerchner, Louis W. Miller, Adam C. Resnick, and Hyman L. Shulman, *The Closed-Loop Planning System for Weapon System Readiness*, Santa Monica, Calif.: RAND Corporation, MG-434-AF, 2006, pp. 5–6.

[50] Unneeded inventory could be generated by catch-up requirements if a safety or pipeline level were generated in error (from obsolete data, for example), or nonoptimized stocks that were above realized demands.

requirements for catch-up and keep-up demands) because the Air Force consistently underfunds DLR buys. We based our funding assumptions on FY 2011–2012 automated budget compilation system (ABCS) data. Thus, some of the "overspending" is really applied to valid requirements that were underfunded in the first place.

The remainder, $90 million in this case, would be unneeded relative to current keep-up demand. Because these are commonly demanded items, that inventory would be drawn down (in about three years, by our estimates) because the inventory management system buys less in subsequent years and ongoing demand generated by FHs eventually "consume" the unneeded inventory.[51] (Our modeling suggests that these parts that exceed immediate demand will be consumed within three years.) If that 10-percent overplanning error is repeated each year, new inventory accumulates as old inventory is consumed, the net result being the carrying of unneeded inventory. The steady-state, or cumulative, level of unneeded inventory generated by 10-percent overplanning is about $150 million.

We applied this relationship between planning error and inventory value to recent historical data. Figure 3.4 shows the value of unneeded inventory (for flying DLRs) that would have accumulated between FYs 2009 and 2015. The solid black line with dots shows the amount of unneeded inventory driven by overplanning (i.e., underflying) purchased in each year. The amount fluctuates year to year in direct proportion to the amount of underflying (hence the spike in FY 2013).

The multicolored areas show the cumulative inventory. Inventory is bought in one year, is consumed (i.e., flown, broken, and repaired until condemned) in subsequent years and thus disappears from inventory. Each year, as new unneeded inventory is purchased and previously unneeded inventory is consumed, the cumulative inventory fluctuates.

In FY 2009, according to our simulation results, about $75 million of unneeded inventory driven by FHs would have been purchased. By FY 2010, more than $30 million would have left the inventory, but $88 million more would have been purchased, creating a cumulative $120 million. By FY 2012, all of the FY 2009 inventory would have been condemned. After that, the cumulative unneeded inventory fluctuates between about $100 million and $150 million.

---

[51] The scope of our analysis is reparable parts, so the parts in question here are not consumed in the traditional sense of consumable parts, which are used until broken and then disposed of. Rather, reparable parts have some finite number of breaks and repairs until they are condemned. Thus, even reparable parts are eventually "consumed."

Figure 3.4. Financial Cost of Unneeded Inventory Resulting from Flying Hour Program Variance

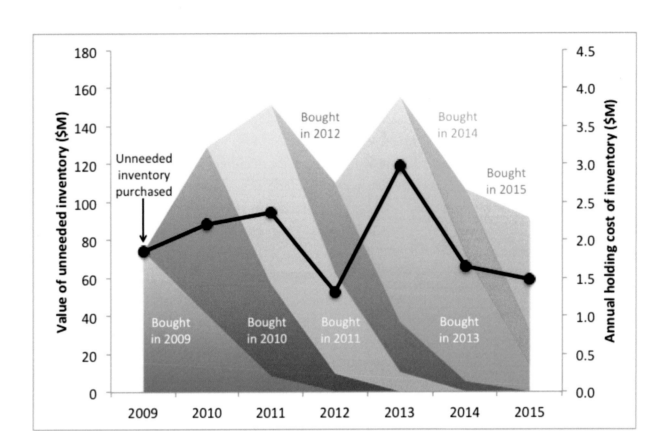

The right-hand y-axis shows the estimated holding cost to the Air Force. We use an approach to estimate holding cost found in Peltz, Cox, et al. (2015).[52] This approach resulted in an annual holding cost of 2.6 percent of the dollar value of the inventory. This is relatively low because we assume that this inventory will not experience obsolescence because these parts are forecast (and therefore often demanded) items. Given all this, we estimate the annual holding cost from unneeded inventory driven by overforecasted FHs to be around $2 million to $4 million per year.

Readiness Effects of Flying Hour Program Variance

Finally, overflying FHs could result in readiness effects, if inventory and repair capacity are insufficient to meet demand. To address this, we looked at a key indicator of supply material availability, mission impaired capability awaiting parts (MICAP) incidents. A MICAP incident

[52] Eric Peltz, Amy G. Cox, Ed Chan, George E. Hart, Daniel Sommerhauser, Caitlin Hawkins, and Kathryn Connor, *Improving DLA Supply Chain Agility: Lead Times, Order Quantities, and Information Flow*, Santa Monica, Calif.: RAND Corporation, RR-822-OSD, 2015.

occurs when a piece of equipment—an aircraft or weapon system, for example—is unable to perform at least one of its missions because it lacks a part that base supply cannot provide.[53]

To analyze the potential effect of FHP variance on MICAPs, we compared the FH data referenced above with historical data on MICAP incidents and duration from FYs 2007 through 2012.[54] This included 3,600 NIINs, applied to 16 MDS, covering more than 605,000 MICAP incidents.

Figure 3.5 shows the results of this comparison. It shows two plots. Each plot shows the relationship of MICAP incidents and duration to FH variance. The top plot shows this for all MICAP cause codes; the bottom plot shows this for cause code H only.[55] On each plot, the purple area shows the number of NIINs with each level of FHP variance (right y-axis). One can see that most of the NIINs fall between –20 percent and +10 percent. These are roughly the same years as the data from Figure 2.1, and the general underplanning bias is consistent.

Also on each plot, the blue line shows the number of MICAP incidents per NIIN in each FHP variance bin, the red line shows the number of MICAP incidents per removal in each bin, and the green line shows the average MICAP duration in each bin. We scaled the number of MICAPs by the number of NIINs and removals in each FHP variance bin to have a scale to compare across NIINs. All of these statistics are then normalized to 100 percent of the value for zero FHP variance. (One can see that all three lines converge at 100 percent on the y-axis at zero FHP variance on the x-axis.)

These plots show no clear or consistent visual trends or relationships between FHP variance and MICAP behavior. Moreover, statistical tests show no statistically significant relationships between FHP variance and MICAP behavior. As we stated above, the relationship between FHs and removals themselves is very noisy. Second, MICAPs happen for all sorts of reasons, many involving supply chain issues having nothing to do with flying activity (e.g., parts obsolescence, lapsed contracts, irregular demands). Thus, there is a baseline of many MICAPs happening on all aircraft types all the time. Finally, in a case of decreasing stock levels because of underforecast demands, the EXPRESS would request additional repair inductions. This is not a panacea—supply support (e.g., consumable-item levels) tends to lag demand changes, but it can help mitigate performance effects.

---

[53] This definition is adapted from Jeremy Arkes and Mary E. Chenoweth, *Estimating the Benefits of the Air Force Purchasing and Supply Chain Management Initiative*, Santa Monica, Calif.: RAND Corporation, MG-584-AF, 2008.

[54] MICAP data were provided by Logistics Directorate, Air Force Sustainment Center (AFSC/LG) on February 4, 2015, and February 26, 2015. Files were drawn from a data warehouse maintained by AFMC, which is populated by monthly pulls or archiving of MICAP data from the Air Force Logistics, Installations, and Mission Support data enterprise.

[55] Cause code H is defined as "Less than full base stock—Stock replenishment requisition exceeds priority group UMMIPS [Uniform Materiel Movement and Issue Priority System] standards. Focus attention on source of supply processing of stock replenishment requisitions." The AFSC/LG personnel who provided the MICAP data informed us that cause code H would be the likeliest to be affected by FHP variance.

Figure 3.5. Effect of Flying Hour Program Variance on Mission Impaired Capability Awaiting Parts Incidence and Duration

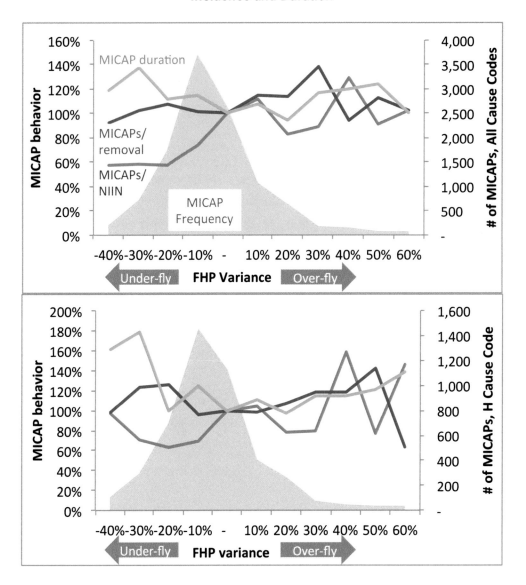

SOURCES: D200 FH data, FYs 2007–2012; MICAP Data Source, FYs 2007–2012.

This does not mean that FH error does not contribute to MICAPs (it likely does, given the small but consistent effect on forecast error)—only that there is no consistent relationship between the two. This finding comes into play when setting expectations for what could be accomplished were FH error significantly reduced. It also does not mean that FH error has no effect on other readiness metrics, such as back orders or customer wait time. We focused here on MICAPs as a kind of proxy for operational readiness.

In the next and final chapter, we discuss conclusions and policy implications.

Chapter Four

Conclusions and Recommendations

Conclusions

Flying Hour Program Variance Has Several Sources

We grouped the causes of FHP variance into three categories. Simple planning error accounts for the basic uncertainties in predicting the FHP in a given year, including the number of pilots, number and type of sorties, sortie duration, and the like. External causes are those that originate outside the service and usually affect the entire enterprise or a significant portion of it, such as contingency operations or congressional action. The third category is internal Air Force decisions, which can cause FHP variance when far-reaching decisions (about force structure, budgets, or FHs themselves) are made after the original FHP is set. Understanding these various sources is key to crafting policy solutions to address and reduce FHP variance.

Flying Hour Program Variance Causes Several Quantifiable, Negative Downstream Effects in the Supply Chain

In this report, we assessed four effects of FHP variance. FHP variance—regardless of its source—increases forecast error, the source of all other downstream effects. Underflying (i.e., overplanning) can incur opportunity cost, leaving money on the table in the budget process. It also incurs financial costs in the form of holding costs for unneeded inventory. Overflying (i.e., underplanning) likely contributes to readiness problems, but our analysis found no statistically significant relationship between FHP variance and MICAPs, one of several important aircraft readiness metrics.

At an Enterprise Level, Most of the Downstream Effects Are Modest

Each effect described above comes with caveats. Though forecast error induced by FHP variance for specific aircraft fleets can be enormous, in most years, the effects on enterprise-level DFA was modest. Except for one year of sequestration, we found the average enterprise-level increase to forecast error to be about five points, or an increase of about 15 percent over baseline error.

Budget opportunity cost can be high—hundreds of millions of dollars in a single year—but many of the recent sources of volatility were events for which flying funding itself was cut from the budget after the FHP was set. Thus, the FH budget was not necessarily too large (i.e., leaving money on the table), even though hours were underflown from their original estimate.

Financial costs incurred from underflying appear to be low, about $2 million to $4 million per year for inventory holding costs in recent years, including the years of sequestration.

Some Individual Programs Do Experience Large Flying Hour Program Variance and Downstream Effects

The AFSC's planning processes do eventually catch large perturbations in FHP inputs, but that does not preclude supply chain planners and operators from having to respond to them. AFSC SMEs reported that, on several occasions, they were caught off guard by seemingly sudden, radical changes to an individual fleet's FH forecast, with little or no communication from planners as to why and with little opportunity to communicate the downstream effects (e.g., canceled contracts, reduced or eliminated repair capacity). A considerable amount of supply chain management time is spent on low-density platforms (which often have unique and expensive parts) to ensure dependable wartime support. This no doubt drives a portion of the AFSC's concern with the accuracy of FHP forecasts.

Some steps have already been taken that might address this gap in communication (at least in part spurred by increased scrutiny from CER). For example, AF/A3 issued a memorandum to increase communication and coordination among HAF organizations involved in the FHP.[56] Besides placing a general emphasis on synchronization and awareness, the memo states that MAJCOMs must explain under- or overexecution and notify AFSC/LG (among other organizations) of approved FH realignment actions.

However, this apparently has not produced the desired results, and the integration of stakeholders involved in FH processes has not been incorporated into efforts aimed at improving FH variance.[57] As a result, AFSC/LG has started an eight-step Cross Command Flying Hour Program Working Group to continue and intensify efforts needed to improve integration and communication regarding the development of FH programs. This should help address extreme program-level perturbations—which appear to be one of AFSC's biggest concerns—that might cause undue FHP variance if supply chain planners are not kept in the loop.

The air staff made several other changes to planning processes: planning FHs at the MDS level, setting improvement targets for FH variance, and including a factor for deployments (which had been excluded in some cases).[58] These changes should be most successful at reducing the natural error in the FH planning process because they are aimed at the fundamental processes that produce the estimates.

Improving Flying Hour Planning Does Not Significantly Improve Spare Part Forecasting

The overarching objective in this analysis, which our specific question about FHP variance addresses, is to achieve CER. Our analysis suggests that there are two separate issues or concerns

---

[56] Giovanni K. Tuck, HQ AF/A3, "FY15 Flying Hour Program Execution Guidance," memorandum, Washington, D.C., October 8, 2014.

[57] Email communication with HQ AFSC personnel on December 19, 2016.

[58] Email communication with HQ AFSC personnel on October 15, 2014.

here. The first is about opportunity cost—essentially, a question of developing a budget. The second is about demand forecasting and its downstream effects.

Significant overplanning can actually reduce budget tradespace by a significant amount that could be invested in other important programs. Thus, improving FH planning could contribute to the accuracy of the POM and free up badly needed resources, in cases in which the driver of overplanning was not a belated cut to the FH budget itself. HAF actions referenced above should help address this.

However, this progress does not necessarily influence the second issue, that of enterprise-level forecast accuracy. Even with a more accurate planning process (i.e., the number-crunching that informs the POM input), FHs remain subject to severe volatility, to all of those features inherent in DoD's current budget system, and to those external events that cannot be anticipated. Given the tenuous nature of the relationships we observed, further efforts to reduce FHP variance might or might not have an *observable* effect on long-term financial cost or readiness because so many other sources of error affect the system. Particular spare parts might be affected more by FHP variance because their removals correlate more highly with flying activity, but any such effects would not be observed system-wide.

The current forecasting system has at least two problems. First, it uses only FHs as a direct, linear input, whereas FHs are themselves volatile.[59] They are subject to the budget process, so the Air Force is pegging its prediction to an input variable that is ever-shifting and based in part on strategy but more so on unforeseeable institutional factors beyond its control and subject to fiscal pressures and budget games.

The second, and maybe more important, feature is that FHs are generally poor predictors of actual removals and repair demands. The individual part level is what matters most for supply chain cost and effectiveness, but there is virtually no correlation with FHs at the part level.[60] Even if numbers of FHs never changed from the original POM forecast, they would still be poor predictors and would give relatively poor DFA and other associated effects. In sum, the Air Force has chosen to drive its parts forecasts for flying DLRs by a single variable that is notoriously volatile and demonstrably unreliable.

So, could DLR removals be better forecast *without* better FHP forecasts? It is beyond the scope of this report to describe a comprehensive approach to improve the Air Force's spare part forecasting system, although we believe that such an approach is needed. However, we did discuss several possibilities with analysts in the 448 SCMW and AFMC/A9A.

---

[59] Manual overrides are used in cases of known or anticipated program changes, but data analysis shows that in aggregate, these overrides generally increase total forecast error.

[60] Past RAND research notes that one fundamental assumption underlying the spare part forecasting system is not supported by the data. In other words, the so-called linearity assumption: "Aircraft failures are driven by a known operational activity: the expected number of failures of a particular part is proportional to a known and measurable quantity, such as flying hours or landing" (see Crawford, 1988, p. v).

The current forecasting system uses a calculated demand rate (removals per FH) and allows for human intervention when equipment specialists have additional applicable information about anticipated future demands (e.g., phasing in or out parts or aircraft). One possibility is, instead of using just a removal rate, supplementing or replacing that with a time-based failure rate, such as demands per quarter. Separate analyses by the PAF research team, 448 SCMW, and AFMC/A9A have shown that DFA can improve when using historical removals instead of the current method of using removal rates, either discounting FHs or ignoring them altogether.[61]

Another possibility is reducing SME intervention in the removal forecast. We found that whether using time-based or FH-based failure rates, on the whole, SME input actually worsened forecasts and reduced DFA. 448 SCMW personnel reported that they briefed personnel at the Office of the Secretary of Defense that DFA was 68.7 percent but would have been 70.2 percent if every item used the system-generated eight-quarter moving average demand rate.[62] This suggests to us that they could achieve a significant increase in DFA (and thus improved readiness and reduced costs) if they merely reverted to this system-generated value.

Even more gains could presumably be garnered with targeted human intervention. The 448 SCMW uses what it calls the DFA team and the Propulsion Analysis and Collaboration for Estimates process to target DFA improvements. They rightly target such changes as time change technical orders and modifications that can (generally) be forecast based on foreseeable events, and such an approach would thus improve on a blanket system-generated moving average demand rate. We simply argue that human intervention, although necessary in some cases, should be used sparingly and only in cases in which it is reliably shown to improve outcomes.

Also, past RAND research also points to some potential solutions. Adams, Abell, and Isaacson (1993) lays out an approach to better forecasting high-demand items using a weighted regression technique.[63] And a number of RAND studies from the mid-1960s showed that sorties rather than FHs drove failures.[64]

Finally, HQ AFSC is working to implement a method called Peak Policy for low-demand, highly variable items. That methodology has been implemented by the Defense Logistics Agency for consumables, and the AFSC is currently working to extend it to include reparables.

---

[61] Some high-demand parts do actually show a reliable relationship between FHs and removals, so the D200 default could be retained. Adams, Abell, and Isaacson (1993) proposed a system that promised even better results.

[62] Email communication with HQ AFSC personnel on May 8, 2017.

[63] Adams, Abell, and Isaacson (1993).

[64] For example, RAND research by William H. McGlothlin, Theodore S. Donaldson, and A. F. Sweetland, as well as Peter J. Francis and Geoffrey B. Shaw, *Effect of Aircraft Age on Maintenance Costs*, Alexandria, Va.: Center for Naval Analyses, CAB D0000289.A2, March 2000.

Recommendations

In light of these findings, we make four recommendations.

Maintain changes to the FHP planning process, which appear to be essentially zero-cost to implement. These changes address mostly our first category of FHP variance, simple planning error, which has driven the majority of the overall volume of enterprise-level FH error in recent years. In addition to providing opportunity cost savings, addressing FHP variance should better balance cost and readiness across Air Force fleets. Resolving recent levels of FHP variance means that the Air Force would not overinvest in one fleet relative to another fleet.[65] However, reliably reducing the cost per unit of readiness (the goal of CER) requires that forecast error be reduced much more significantly than reducing FHP variance alone can accomplish.

Second, **continue to support and extend efforts to improve integration and communication** across commands and between the operational and supply chain communities, such as the Cross Command Flying Hour Program Working Group started by AFSC/LG. This type of effort seems like the best hope to address our third category of FHP variance, internal Air Force decisions. The communication and coordination inherent in something like the working group can help avoid surprises or potentially mitigate shortsighted decisions.

Third, **consider management mechanisms that could dampen the downstream volatility caused by FHP variance**. One approach to this would have supply chain managers explicitly incorporate uncertainty and total cost into their decision calculus. Sometimes downstream supply chain decisions are incremental (e.g., determining the number of spare parts to buy or the number of maintainers to hire). Thus, more FHs means buying more parts, and more overflying, for example, means higher inventory costs for unneeded parts. But sometimes decisions are essentially binary (e.g., starting or stopping a repair line altogether, or canceling or letting a contract). There could be cases in which understanding and sharing information about these thresholds (which only supply chain specialists would know) could change the decision calculus of budget planners, if only enough to avoid crossing a potentially costly threshold.

In cases for which FHs depart drastically from history (assuming that history is somewhat consistent), one reasonable question is this: Is the cost (financial or otherwise) of executing this action according to stated requirements, and then reversing it, greater or lesser than if historical requirements are used? Suppose that a requirement is provided that is half the number of historical FHs for an aircraft fleet. It is possible that the cost of canceling a spare part buy contract, then having to pay to restart it and mitigate parts shortages in the meantime, is greater than the cost of simply maintaining the contract and holding slightly more inventory until the numbers of FHs return to historically typical levels (assuming that the FHs were not simply following a reduction in force structure).

---

[65] One can imagine a cynical planner deliberately overestimating FHs for a particular MDS and, in the year of execution, simply achieving higher mission capability rates or fuller spares kits when those FHs do not materialize, to the detriment of other MDS that were more conservative in their planning.

For small programs, perhaps the costs of overplanning (holding a little more inventory) are less than underplanning (experiencing and frantically remedying readiness problems). For larger programs, especially those with more-robust supply chains (e.g., commercial equivalent or derivative aircraft, or aircraft with global sales), the threshold might be lower because there are more options to accommodate a shortfall.

This approach also goes hand-in-hand with better communication. If up- and downstream personnel communicate clearly and regularly about the real-life dynamics and uncertainties in shaping and reshaping the FHP and the thresholds for downstream decisions, they could collectively reach decisions that reflect the least cost to the Air Force. This could result in adjusting FHP requirements, downstream decisions, or both. But it is possible only in an environment with healthy, robust communication. Some of this communication already happens downstream because equipment specialists and item managers communicate about demand rates and demand plans. Perhaps that communication could be broadened to include upstream planners and demanders to better understand costs and benefits.

Having flexibility in a system—any system—costs something. FH funding can be easier to obtain, has more leverage than other programs, and is thus a useful tool in the bureaucratic maneuvering that takes place every year. Having a budgeting system that allows upstream planners to change the FHP as funding becomes available to provide readiness is a very valuable capability. Until the nature of the funding process for FHs radically changes, that will remain a fact of life. But that flexibility comes at a cost. Upstream decisionmakers should be aware of that cost and the effort required to keep supporting the critical readiness they desire.

Another approach to dampening downstream volatility is to use a more automated strategy. In the current system, the computational process for peacetime spares requirements recomputes the demand rate and resupply times each quarter without regard to previous quarters' values. This approach can induce some degree of volatility because the rates and pipeline times shift. A different strategy would be to update these values only when there has been a statistically supportable change in the mean value.

Fourth, **continue to take a holistic approach to improving spare part forecasts**—i.e., look beyond the FHP—and focus improvement efforts on issues that are the largest sources of forecast variance. Further efforts to improve forecasting should focus on (admittedly harder) problems, such as the forecasting algorithms themselves, inventory policies (such as Peak Policy), and the information systems that contain them.

We understand that the Air Force sought to improve its spare part forecasting system with the failed implementation of Expeditionary Combat Support System, and the Air Force is again investigating information system solutions to this (and other) issues. As new systems come online, one key question is when FHs should be used for forecasts. Empirical analyses can be performed to assess which items have a strong enough correlation to be useful or, in other cases, where thresholds should be set such that FHP variance beyond a certain point would trigger some action. To the degree that these new systems provide insights into these questions and the ability to better calibrate spare part decisionmaking, the Air Force can realize some long-awaited benefits to readiness and cost-effectiveness.

Simulation Tool for DLR Decisions

This appendix explains more about the computer simulation we used to estimate opportunity costs and unneeded inventory in Chapter Three. We begin by listing the key sources and assumptions for simulation (Table A.1).

Table A.1. Key Sources and Assumptions for Simulation

Data source	Data inputs
ABCS (FY11, FY12)	Buy/repair split
	Amount of total requirement (buy and repair) that is funded
D200	Differences between planned and actual demands
	Buy vs. repair costs
AFTOC	DLR costs of FHP (includes unit-level removals, not depot-level removals)

NOTE: AFTOC = Air Force Total Ownership Cost.

Given planned and actual numbers of FHs, our simulation computes removals, condemnations, carcasses, buys, and repairs. For each year, the simulation models planning and execution. In planning, the model calculates the total funded requirement, which is equal to the funded level of additives (including safety stock and pipeline levels), plus the expected number of breaks predicted based on planned numbers of FHs and historical breaks per FH, minus the numbers of any serviceable assets on hand. We assume that the system would prefer to satisfy this requirement through repair. The planned repair is equal to the lesser of the requirement, or the number of carcasses available for repair, minus the number of any planned condemnations. Buys (both planned and actual) are equal to whatever portion of the funded requirement cannot be satisfied though repair. At the end of the planning phase, the number of serviceable assets is increased according to the buys.

In execution, we simulate a true number of breaks for each NIIN, which might differ from the expected number of breaks. The difference between the numbers of expected and actual breaks (prediction error) is drawn from a distribution based on data from D200. Similar to what is done for the planned requirement, we calculate an execution requirement, which is the sum of the number of funded additives and actual breaks, minus the numbers of serviceable assets. We assume that the number of repairs will be equal to the minimum of the number of carcasses available (after condemnations in execution) and the execution requirement. At the end of execution, the number of serviceable assets and the number of carcasses are updated to reflect the breaks and repairs that occurred in execution.

We assume that, under perfect planning, the funded requirement is equal to the keep-up requirement, which funds the buys and repairs for the breaks expected during the fiscal year. We use this assumption to calibrate the funded additive levels in the model (A_i), such that, when $pFH_i^n = aFH_i^n$ for all i and all n, the funding levels for buy or repair match those in ABCS for FYs 2011 and 2012. Table A.2 lists the variables for simulation.

Table A.2. Description of Variables for Simulation

Variable Name	Description
i	Item (DLR) index
pFH_i^n	Number of planned FHs for item i in iteration n
$aFHi_i^n$	Number of actual FHs for item i in iteration n
A_i	Funding level for additives for item i. Funding level is relative to the full (and partly unfunded) additive requirement. Funding level −3 would indicate that the funded additive level is three less than the full additive requirement
f_i	Number of breaks per FH rate for item i
pBr_i^n	Expected (planned) number of breaks for item i in iteration n
pR_i^n	Expected (planned) number of repairs for item i in iteration n
aBr_i^n	Actual number of breaks for item i in iteration n
aR_i^n	Actual number of repairs for item i in iteration n
S_i^n	Number of serviceable items i in iteration n
Bu_i^n	Number of buys of item i in iteration n
$pReq_i^n$	Planned requirement for item i in iteration n
$aReq_i^n$	Actual (execution) requirement for item i in iteration n
$pCon_i^n$	Planned number of condemnations for item i in iteration n
$aCon_i^n$	Actual number of condemnations (in execution) for item i in iteration n
Car_i^n	Number of carcasses of item i in iteration n

Figure A.1 shows how the simulation models the planning process. We assume that the total planned requirement, which could be satisfied through either buy or repair, is equal to:

$$pReq_i^n = A_i + pBr_i^n - S_i^n,$$

which is the funded level of additives (including safety stock and pipeline requirements), plus the expected number of DLR breaks, which is equal to the planned number of FHs times the historical breaks–per–FH rate from D200 ($pBr_i^n = f_i * pFH_i^n$), minus the number of serviceable assets on hand.

Figure A.1. How the Simulation Model Accounts for the Planning Process

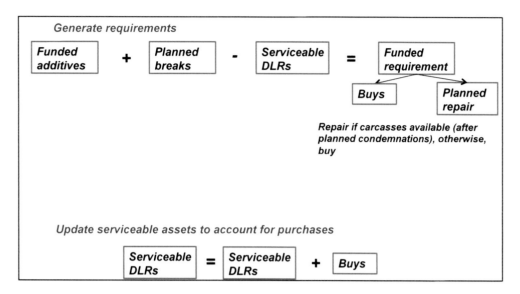

We further assume that the system would prefer to satisfy the requirement through repair than through purchasing new DLRs. The number of planned repairs is given by

$$pR_i^n = \min(pReq_i^n, Car_i^n - pCon_i^n).$$

We assume a condemnation rate of 13 percent per repair, which we determined through calibration to match the observed buy-or-repair split from ABCS (higher condemnation rates increase the proportion of the requirement that is fulfilled through buys). The number of buys (both planned and actual), then, is equal to the funded DLR requirement minus the planned repair:

$$Bu_i^n = pReq_i^n - aR_i^n.$$

The final step in modeling DLR planning is to update the number of serviceable assets to include the newly purchased DLRs; we call this intermediate number of serviceables $S_i^{n+\frac{1}{2}}$; it is given by

$$S_i^{n+\frac{1}{2}} = S_i^n + Bu_i^n.$$

Figure A.2 shows how we model execution in the simulation. We first compute the number of breaks, which is drawn from a distribution that is conditional on the expected number of breaks. The distribution is based on historical D200 data for FH-driven items from FYs 2010 through 2012. We then determine the number of condemnations, which is drawn from a Poisson distribution based on the number of breaks and the condemnation (per break) rate. We then update the number of carcasses available (we call the number of carcasses available at this

39

Figure A.2. How the Simulation Model Accounts for the Execution Process

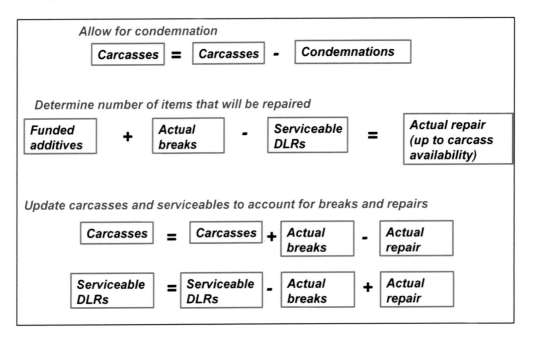

intermediate step $Car_i^{n+\frac{1}{2}}$) for repair by subtracting the condemnations from the number of available carcasses:

$$Car_i^{n+\frac{1}{2}} = Car_i^n - aCon_i^n.$$

We then determine the execution requirement, which is given by

$$aReq_i^n = A_i + aBr_i^n - S_i^{n+\frac{1}{2}},$$

equal to the funded level of additives, plus the number of breaks, minus the number of serviceable DLRs on hand. The number of repairs done in execution is the minimum of the number of available carcasses and execution repair requirement; this is given by

$$aR_i^n = \min(Car_i^{n+\frac{1}{2}}, aReq_i^n).$$

The final step in simulating execution is to update the number of carcasses and the number of serviceable DLRs. The number of carcasses increases with DLR breaks and decreases with repairs; the number of DLRs increases with repairs and decreases with breaks. These quantities are given by

$$S_i^{n+1} = S_i^{n+\frac{1}{2}} + aR_i^n - aBr_i^n$$
$$Car_i^{n+1} = Car_i^{n+\frac{1}{2}} - aR_i^n + aBr_i^n.$$

40

Figure A.3 shows the DLR buy spending for a single year (in billions of FY 2013 dollars), as computed by our simulation. The middle column shows a spending level of $800 million with perfect FHP prediction. The error bars show the level of uncertainty in buy requirement because of fluctuating demands.

Underplanning reduced DLR buy spending by about $200 million, and overplanning increased spending by about $250 million. Overplanning costs more than underplanning saves because, relative to a steady state, overplanning increases the mix of demands that must be fulfilled with purchases rather than repairs. Underplanning represents a greater mix of repairs, rather than purchases, that were avoided. Because repairs are less costly than purchases, underplanning results in less savings than overplanning costs.

Furthermore, overplanning has more long-term financial effects because overbought parts must be reduced by attrition, but underbuying can be reversed in the next planning cycle (albeit with potential readiness effects in the intervening period).

Figure A.4 shows the same results for repairs.

Underplanning reduced repair spending by about $180 million, whereas overplanning increased spending by about $170 million. Furthermore, overplanning has more long-term financial effects because overplanned capacity might be partially unusable, and therefore unrecoverable, whereas underplanning capacity can usually be remedied by working overtime. Having estimated these error–cost relationships with our simulation, we can now apply them to historical levels of FHP variance.

Figure A.3. Spare Part Buy Costs of Planning Error

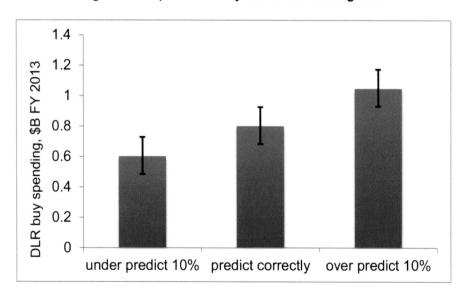

Figure A.4. Repair Costs of Planning Error

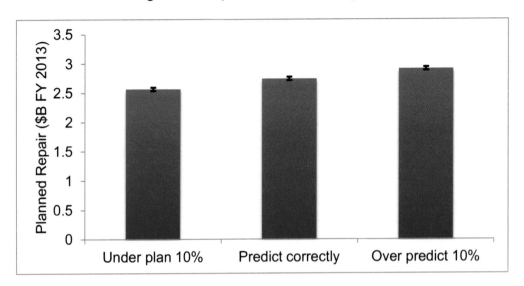

Appendix B
Additional Data on Flying Hours and Removals

This appendix shows data plots in addition to those in Chapter Three. Figures B.1–B.4 show FH data for fixed-wing aircraft types for FYs 2008–2011, respectively. Each diamond represents one MDS, showing the actual FHs (x-axis) compared with the FH variance as a percentage of forecast FHs (y-axis). The further away a diamond is from the centerline, the greater the variance from what was planned. A diamond above the line indicates that the MDS flew more hours than were budgeted; below the line, fewer hours. We have truncated the y-axis at 100-percent overflying.

Figure B.1. FY 2008 Aircraft-Type Flying Hour Error

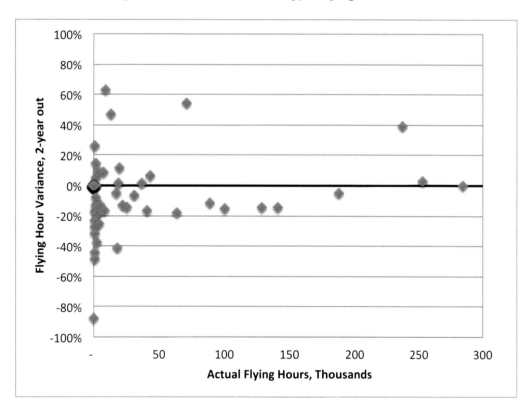

SOURCE: D200, 2011 (two-year-out forecast).

43

Figure B.2. FY 2009 Aircraft-Type Flying Hour Error

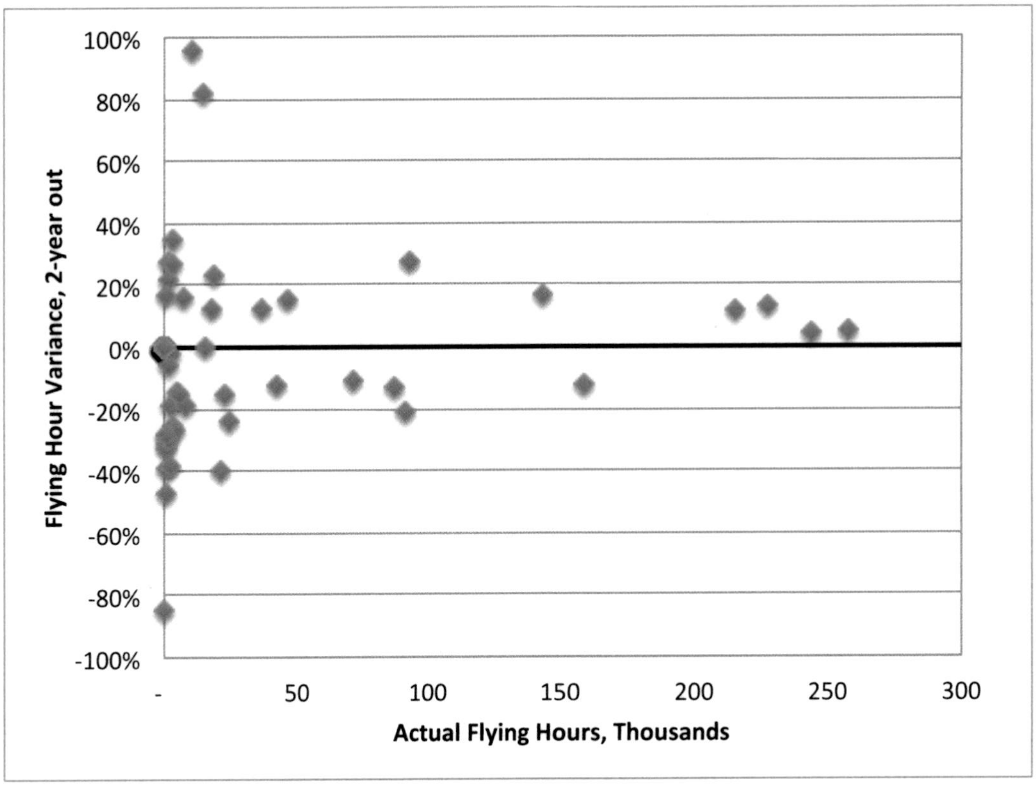

SOURCE: D200, 2011 (two-year-out forecast).

Figure B.3. FY 2010 Aircraft-Type Flying Hour Error

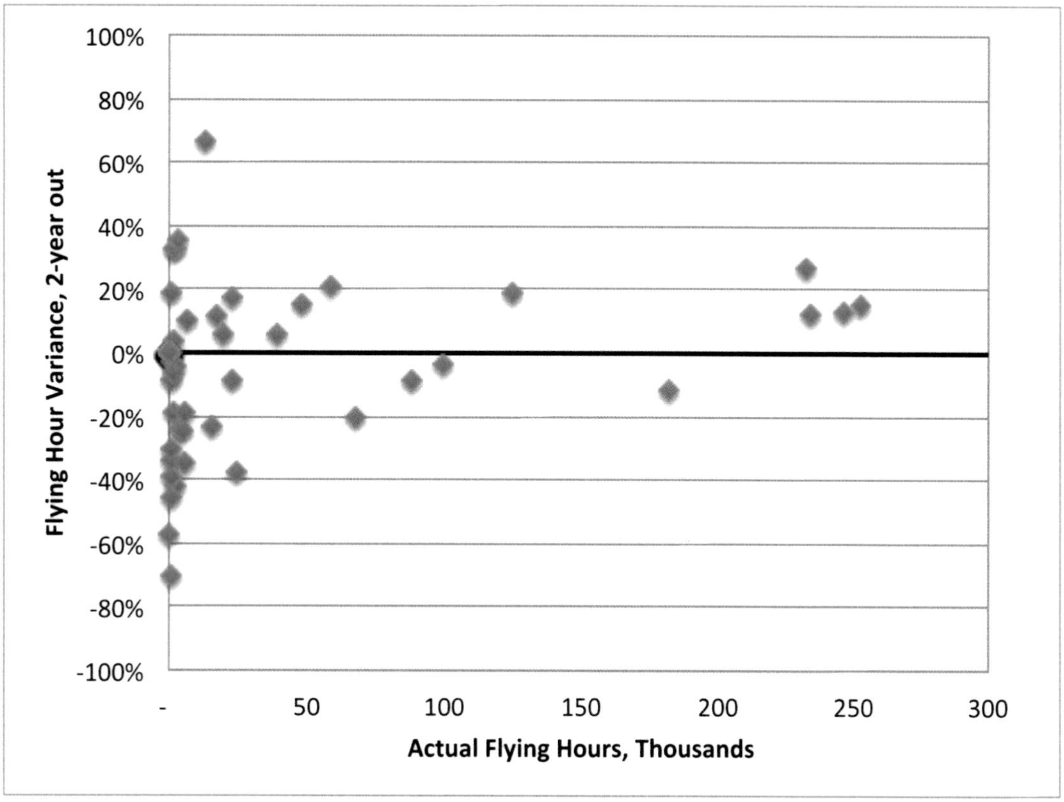

SOURCE: D200, 2011 (two-year-out forecast).

Figure B.4. FY 2011 Aircraft-Type Flying Hour Error

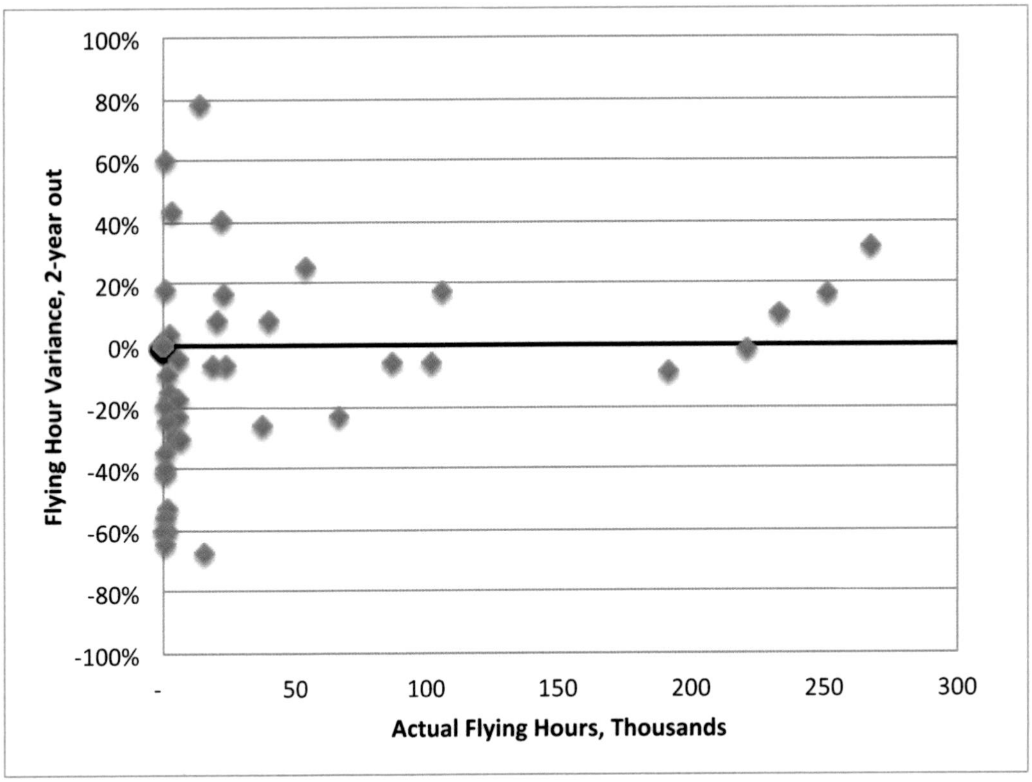

SOURCE: D200, 2011 (two-year-out forecast).

References

Abell, John B., *Estimating Requirements for Aircraft Recoverable Spares and Depot Repair: Executive Summary*, Santa Monica, Calif.: RAND Corporation, R-4215-AF, 1993. As of July 28, 2017:
http://www.rand.org/pubs/reports/R4215.html

Adams, John L., John B. Abell, and Karen E. Isaacson, *Modeling and Forecasting the Demand for Aircraft Recoverable Spare Parts*, Santa Monica, Calif.: RAND Corporation, R-4211-AF/OSD, 1993. As of July 28, 2017:
http://www.rand.org/pubs/reports/R4211.html

Arkes, Jeremy, and Mary E. Chenoweth, *Estimating the Benefits of the Air Force Purchasing and Supply Chain Management Initiative*, Santa Monica, Calif.: RAND Corporation, MG-584-AF, 2008. As of July 28, 2017:
http://www.rand.org/pubs/monographs/MG584.html

Boito, Michael, Thomas Light, Patrick Mills, and Laura H. Baldwin, *Managing U.S. Air Force Aircraft Operating and Support Costs: Insights from Recent RAND Analysis and Opportunities for the Future*, Santa Monica, Calif.: RAND Corporation, RR-1077-AF, 2016. As of July 28, 2017:
http://www.rand.org/pubs/research_reports/RR1077.html

Brown, Bernice B., *Characteristics of Demand for Aircraft Spare Parts*, Santa Monica, Calif.: RAND Corporation, R-292, 1956. As of July 28, 2017:
http://www.rand.org/pubs/reports/R292.html

Brown, Bernice B., and Murray A. Geisler, *Analysis of the Demand Patterns for B-47 Airframe Parts at Air Base Level*, Santa Monica, Calif.: RAND Corporation, RM-1297, 1954. As of July 28, 2017:
http://www.rand.org/pubs/research_memoranda/RM1297.html

Cohen, Irv K., John B. Abell, and Thomas F. Lippiatt, *Coupling Logistics to Operations to Meet Uncertainty and the Threat (CLOUT): An Overview*, Santa Monica, Calif.: RAND Corporation, R-3979-AF, 1991. As of July 30, 2017:
http://www.rand.org/pubs/reports/R3979.html

Crawford, Gordon B., *Variability in the Demands for Aircraft Spare Parts: Its Magnitude and Implications*, Santa Monica, Calif.: RAND Corporation, R-3318-AF, 1988. As of July 28, 2017:
http://www.rand.org/pubs/reports/R3318.html

Francis, Peter J., and Geoffrey B. Shaw, *Effect of Aircraft Age on Maintenance Costs*, Alexandria, Va.: Center for Naval Analyses, CAB D0000289.A2, March 2000.

Goldfein, David L., "Department of the Air Force Presentation to the Subcommittee on Readiness, United States House of Representatives Committee on Armed Services," February 12, 2016. As of December 22, 2017:
http://docs.house.gov/meetings/AS/AS03/20160212/104347/
HHRG-114-AS03-Wstate-GoldfeinD-20160212.pdf

Hess, Tyler, *Cost Forecasting Models for the Air Force Flying Hour Program*, Wright-Patterson Air Force Base, Ohio: Air Force Institute of Technology, March 2009.

Hillestad, Richard, Robert Kerchner, Louis W. Miller, Adam C. Resnick, and Hyman L. Shulman, *The Closed-Loop Planning System for Weapon System Readiness*, Santa Monica, Calif.: RAND Corporation, MG-434-AF, 2006. As of July 28, 2017:
http://www.rand.org/pubs/monographs/MG434.html

Hodges, James S., and Raymond A. Pyles, *Onward Through the Fog: Uncertainty and Management Adaptation in Systems Analysis and Design*, Santa Monica, Calif.: RAND Corporation, R-3760-AF/A/OSD, 1990. As of July 28, 2017:
http://www.rand.org/pubs/reports/R3760.html

Moore, Josh, *448th Supply Chain Management Wing Demand Forecast Accuracy Sep 08–Mar 14 Briefing*, October 2014.

Moore, Josh, *448th Supply Chain Management Wing FY14 DFA Results Briefing*, January 14, 2015.

Muckstadt, John A., *Analysis and Algorithms for Service Parts Supply Chains*, New York: Springer, 2005.

Peltz, Eric, Amy G. Cox, Ed Chan, George E. Hart, Daniel Sommerhauser, Caitlin Hawkins, and Kathryn Connor, *Improving DLA Supply Chain Agility: Lead Times, Order Quantities, and Information Flow*, Santa Monica, Calif.: RAND Corporation, RR-822-OSD, 2015. As of July 28, 2017:
http://www.rand.org/pubs/research_reports/RR822.html

Pyles, Raymond A., *The Dyna-METRIC Readiness Assessment Model: Motivation, Capabilities, and Use*, Santa Monica, Calif.: RAND Corporation, R-2886-AF, 1984. As of July 28, 2017:
http://www.rand.org/pubs/reports/R2886.html

Sherbrooke, Craig C., *METRIC: A Multi-Echelon Technique for Recoverable Item Control*, Santa Monica, Calif.: RAND Corporation, RM-5078-PR, 1966. As of July 28, 2017:
http://www.rand.org/pubs/research_memoranda/RM5078.html

———, "METRIC: A Multi-Echelon Technique for Recoverable Item Control," *Operations Research*, Vol. 16, No. 1, 1968, pp. 122–141.

Slay, F. Michael, Tovey C. Bachman, Robert C. Kline, T. J. O'Malley, Frank L. Eichorn, and Randall M. King, *Optimizing Spares Support: The Aircraft Sustainability Model*, McLean, Va.: Logistics Management Institute, AF501MR1, 1996.

Tuck, Giovanni K., HQ AF/A3, "FY15 Flying Hour Program Execution Guidance," memorandum, Washington, D.C., October 8, 2014.

U.S. Air Force, *Strategic Planning System*, Air Force Policy Directive 90-11, March 26, 2009.

———, *Flying Hour Program Management*, Air Force Instruction 11-102, August 30, 2011.

———, *Cost Effective Readiness LOE #2: Flying Hour Program Inputs*, briefing, 2013a.

———, *2012–2022 Air Force Enterprise Logistics Strategy*, Version FY14.2, October 16, 2013b.

U.S. General Accounting Office, *Observations on the Air Force Flying Hour Program*, Washington, D.C., NSIAD-99-165, July 8, 1999.

U.S. Government Accountability Office, *Defense Inventory: Actions Needed to Improve the Defense Logistics Agency's Inventory Management*, Washington, D.C., GAO-14-495, June 2014.